Dieter Leseberg

Strict Topological Extensions and Power-objects for b-convergence

Dieter Leseberg

Strict Topological Extensions and Power-objects for b-convergence

LAP LAMBERT Academic Publishing

Impressum / Imprint
Bibliografische Information der Deutschen Nationalbibliothek: Die Deutsche Nationalbibliothek verzeichnet diese Publikation in der Deutschen Nationalbibliografie; detaillierte bibliografische Daten sind im Internet über http://dnb.d-nb.de abrufbar.

Bibliographic information published by the Deutsche Nationalbibliothek: The Deutsche Nationalbibliothek lists this publication in the Deutsche Nationalbibliografie; detailed bibliographic data are available in the Internet at http://dnb.d-nb.de.

Coverbild / Cover image: www.ingimage.com

Verlag / Publisher:
LAP LAMBERT Academic Publishing
ist ein Imprint der / is a trademark of
OmniScriptum GmbH & Co. KG
Heinrich-Böcking-Str. 6-8, 66121 Saarbrücken, Deutschland / Germany
Email: info@lap-publishing.com

Herstellung: siehe letzte Seite /
Printed at: see last page
ISBN: 978-3-659-51205-6

We consider *antiform* b-convergence as a *common* generalization of *preuniform convergence*, well-known *point-convergences* and suitable *set-convergences* as well.
The corresponding defined category ***ab-CONV*** is a topological construct which is cartesian closed. Then the above mentioned categories can be nicely embedding into ***ab-CONV***. Moreover, we will establish a one-to-one correspondence between some b-convergence and the related symmetric strict topological extension.

MSC 2010: 54A20 54B30 54C35 54D35 54E15
Keywords and phrases: Bounded topology; antiform b-convergence; cartesian closed topological category; strict topological extension;

Contents

1 Introduction

The new created concept "Bounded Topology", a convenient foundation for Topology (Leseberg 2009) deals with a lot kind of structures examined by topologists in the past: Especially *generalized convergence spaces*, now defined as b-converges (Leseberg 2008) appear in a new context with the same of its convenient properties like being a trong topological universe (Preuß2002). So it was proved by Leseberg that the category of pointed b-convergence spaces forms a strong topological universe in which the classical constructs ***TOP*** and ***UNIF*** can both be embedded in particular nice fashion. Moreover, the category ***PUCONV*** of preuniform convergence spaces and related maps are being involved as well. Hence ***SUCONV*** the category of semiuniform spaces can be also considered as special subcategory. Since the structure of simple convergence (= poinwise convergence), uniform convergence and continuous convergence can be derived from the natural function space structure in ***SUCONV*** (Preuß2002) we study in 6. function space structures in the construct ***ab-CONV*** and in suitable subconstructs.

Additionally, we will stablish a ertain kind of *symmetric topological extension*, which is in one-to-one correspondence to some related b-convergence. Thus, among others the Doîtchînov-extension (1985) now also can be considered as special case of this *general* concept.

2 Basic concepts and notations

As usual $\underline{P}X$ denotes the power set of a set X, and we use $\mathcal{B}^X \subset \underline{P}X$ to denote a *collection* of *bounded* subsets of X, known as $\underline{B}$-set, [14] or *boundedness* on X, respectively, i. e. $\mathcal{B}^X$ has the following properties:

(b_0) $\emptyset \in \mathcal{B}^X$;

(b$_1$) $B_2 \subset B_1 \in \mathcal{B}^X$ imply $B_2 \in \mathcal{B}^X$;

(b$_2$) $x \in X$ implies $\{x\} \in \mathcal{B}^X$.

Then for $\underline{B}$-sets $\mathcal{B}^X, \mathcal{B}^Y$ a function $f : X \longrightarrow Y$ is called *bounded* iff it satisfies (b) $\{f[B] : B \in \mathcal{B}^X\} \subset \mathcal{B}^Y$. We denote by ***BOUND*** the corresponding category and mention here its convenient property for being a *topological universe* (see Section 4.). Here, we point out that following $\underline{B}$-sets are playing some *important* role:
Firstly, let us consider the so-called *discrete boundedness* $\mathcal{D}^X$ defined by setting:

$$\mathcal{D}^X := \{\emptyset\} \cup \{\{x\} : x \in X\};$$

secondly, a boundedness $\mathcal{B}^X$ is called *saturated* iff it satisfies

(sat) $X \in \mathcal{B}^X$, hence $\mathcal{B}^X$ equals $\underline{P}X$;

thirdly, let us consider the *finite* boundedness $\mathcal{B}^X_{fin}$ by setting $\mathcal{B}^X_{fin} := \{B \subset X : B \text{ is finite }\}$, and at last for $B \subset X$ we always have the *subboundedness* $B^{\downarrow} := \{\{x\} : x \in X\} \cup \{B_1 \subset X : B_1 \subset B\}$. By adding suitable properties for a given boundedness one may obtain some specific $\underline{B}$-sets like *bornologies* etc., [7].

Definition 2.1. For a set X we call a pair $(\mathcal{B}^X, \tau)$ consisting of a boundedness $\mathcal{B}^X$ and *b-convergence operator* (on $\mathcal{B}^X$) $\tau : \mathcal{B}^X \longrightarrow \underline{P}(\boldsymbol{FIL}(X \times X))$ a *b-convergence* (on X), and the triple $(X, \mathcal{B}^X, \tau)$ a *b-convergence space* iff the following axioms are satisfied:

(bc$_1$) $B \in \mathcal{B}^X, \mathcal{U} \in \tau(B)$ and $\mathcal{U} \subset \mathcal{V} \in \boldsymbol{FIL}(X \times X)$ imply $\mathcal{V} \in \tau(B)$;

(bc$_2$) $\tau(\emptyset) := \{\underline{P}(X \times X)\}$;

(bc$_3$) $x \in X$ implies $\dot{x} \times \dot{x} \in \tau(\{x\})$.

(Here, $\boldsymbol{FIL}(X \times X)$ denotes the set of all *uniform* filters on X, including the *nullfilter*. $\overset{\bullet}{x}$ denotes the filter generated by the set $\{x\}$. In general, for filters $\mathcal{F}, \mathcal{G} \in \boldsymbol{FIL}(X)$ their *crossproduct* is defined by

$$\mathcal{F} \times \mathcal{G} := \{R \subset X \times X : \exists F \in \mathcal{F} \exists G \in \mathcal{G} R \supset F \times G\}.$$

If $\mathcal{U} \in \tau(B)$ for some $B \in \mathcal{B}^X$, we say the uniform filter $\mathcal{U}$ *b-converges* to B. Given two b-convergence spaces $(X, \mathcal{B}^X, \tau_X)$, $(Y, \mathcal{B}^Y, \tau_Y)$, a function $f : X \longrightarrow Y$ is called *b-uniformly continuous* iff it is bounded, and in addition we have that f *preserves* uniform filters in the sense that (buc) $B \in \mathcal{B}^X \backslash \{\varnothing\}$ and $\mathcal{U} \in \tau_X(B)$ imply $(f \times f)(\mathcal{U}) \in \tau_Y(f[B])$, where $(f \times f)(\mathcal{U}) := \{V \subset Y \times Y : \exists U \in \mathcal{U}\, V \supset (f \times f)[U]\}$ with $(f \times f)[U] := \{(f \times f)(x, z) : (x, z) \in U\} = \{(f(x), f(z)) : (x, z) \in U\}$ holds. Moreover, we denote the corresponding construct by b-$\boldsymbol{CONV}$ and mention here its interesting property for being a topological category (see Section 4).

Simple examples 2.2. Let $\mathcal{B}^X$ be boundedness on a set X, then we consider the following b-convergence operators on $\mathcal{B}^X$, i. e.

(i) $\tau_1(\emptyset) := \{\underline{P}(X \times X)\}$ and $\tau_1(B) := \{\mathcal{U} \in \boldsymbol{FIL}(X \times X) : \exists x \in B \sec \mathcal{U} \subset \overset{\bullet}{x} \times \overset{\bullet}{x}\}$, otherwise, (Here, $\sec \mathcal{U} := \{R \subset X \times X : \forall U \in \mathcal{U}\, U \cap R \neq \emptyset\}$);

(ii) $\tau_2(\emptyset) := \tau_1(\emptyset)$ and $\tau_2(B) := \{\mathcal{U} \in \boldsymbol{FIL}(X \times X) : \forall x \in B \sec \mathcal{U} \subset \overset{\bullet}{x} \times \overset{\bullet}{x}\}$, otherwise;

(iii) $\tau_3(\emptyset) := \tau_2(\emptyset)$ and $\tau_3(B) := \{\mathcal{U} \in \boldsymbol{FIL}(X) : \overset{\bullet}{B} \times \overset{\bullet}{B} \subset \mathcal{U}\}$, otherwise, (Here, $\overset{\bullet}{B} := \{T \subset X : B \subset T\}$;

(iv) $\tau_4(\emptyset) := \tau_3(\emptyset)$ and $\tau_4(B) := \{\mathcal{U} \in \boldsymbol{FIL}(X) : \exists \mathcal{F} \in \boldsymbol{FIL}(X)\, \overset{\bullet}{B} \times F \subset \mathcal{U}\}$, otherwise;

(v) $\tau_5(\emptyset) := \tau_4(\emptyset)$ and $\tau_5(B) := \{\mathcal{U} \in \boldsymbol{FIL}(X) : \exists \mathcal{F} \in \boldsymbol{FIL}(X)\, \mathcal{F} \times \mathcal{F} \subset \mathcal{U}\}$, otherwise.

Remark 2.3. With respect to the given boundedness $\mathcal{B}^X$ we call a b-convergence space $(X, \mathcal{B}^X, \tau)$ *discrete* or *saturated*, respectively if $\mathcal{B}^X$ possesses the *corresponding* property. Moreover, we denote by ***DISb-CONV*** and ***SATb-CONV*** the related full subcategories of ***b-CONV***. In addition we will now present some *fundamental* types of b-convergences or spaces, respectively.

Definition 2.4. For a boundedness $\mathcal{B}^X$ a b-convergence operator τ is called

(1) *isoform* iff it satisfies

  (i) $\emptyset \neq B_1 \subset B_2 \in \mathcal{B}^X$ imply $\tau(B_1) \subset \tau(B_2)$;

(2) *antiform* iff it satisfies

  (a) $\emptyset \neq B_1 \subset B_2 \in \mathcal{B}^X$ imply $\tau(B_2) \subset \tau(B_1)$;

(3) *equiform* iff it satisfies

  (e) $B_1 \neq \emptyset \neq B_2$ imply $\tau(B_1) = \tau(B_2)$.

By ***ib-CONV***, ***ab-CONV*** and ***eb-CONV***, respectively we denote the corresponding full subcategories of ***b-CONV***.

Remark 2.5. As easily seen, the b-convergence operators τ_1, τ_3 and τ_4 are isoform. τ_5 is equiform, whereas τ_2 is antiform.

Structural examples 2.6. (i) Let (X,J) be a *preuniform* convergence space, sec. 9. For a boundedness $\mathcal{B}^X$ we set for $B \in \mathcal{B}^X \setminus \{\varnothing\} \tau_J(B) := J$ and $\tau_J(\emptyset) := \{\underline{P}(X \times X)\}$;

(ii) For a *set-convergence* space $(X, \mathcal{B}^X, q)$, sec 9. we put

$\tau_q(\emptyset) := \{\underline{P}(X \times X)\}$ and

$\tau_q(B) := \{\mathcal{U} \in \boldsymbol{FIL} : \exists \mathcal{F} \in \boldsymbol{FIL}(X)(\mathcal{F} q B \text{ and } \mathcal{F} \times \mathcal{F} \subset \mathcal{U})\}$, otherwise;

(iii) Let (X, r) be a *filtermerotopic* space (=filterspace), sec 9. For a boundedness $\mathcal{B}^X$ we set for $B \in \mathcal{B}^X \backslash \{\varnothing\}$

$$\tau_r(B) := \{\mathcal{U} \in \boldsymbol{FIL}(X \times X) : \exists \mathcal{F} \in r\mathcal{F} \times \mathcal{F} \subset \mathcal{U}\}$$

$$\text{and } \tau_r(\emptyset) := \{\underline{P}(X \times X)\};$$

(iv) Let (X, p) be *generalized convergence* space, sec 9. For a boundedness $\mathcal{B}^X$ we set for $B \in \mathcal{B}^X \backslash \{\varnothing\}$

$$\tau_p(B) := \{\mathcal{U} \in \boldsymbol{FIL}(X \times X) : \forall x \in B \exists \mathcal{F} \in \boldsymbol{FIL}(X)(\mathcal{F}px$$

$$\text{and } \mathcal{F} \times \mathcal{F} \subset \mathcal{U})\} \text{ and } \tau_p(\emptyset) := \{\underline{P}(X \times X)\}.$$

Definition 2.7. For a boundedness $\mathcal{B}^X$ a b-convergence operator τ is called

(1) *set-pointed* iff it satisfies

(sp) $B \in \mathcal{B}^X$ implies $\dot{B} \times \dot{B} \in \tau(B)$;

an isoform b-convergence operator τ is called

(2) *pointed* iff it satisfies

(p) $B \in \mathcal{B}^X \backslash \{\varnothing\}$ implies $\tau(B) \subset \cup\{\tau(\{x\}) : x \in B\}$;

an antiform b-convergence operator τ is called

(3) *repointed* iff it satisfies

(rp) $B \in \mathcal{B}^X \backslash \{\varnothing\}$ implies $\cap\{\tau(\{x\}) : x \in B\} \subset \tau(B)$.

By ***SET b-CONV***, ***pb-CONV*** and ***rpb-CONV*** we denote the corresponding full subcategories of ***b-CONV***.

Remark 2.8. Note, that for pointed b-convergence spaces $(X, \mathcal{B}^X, \tau)$ we always have for each $B \in \mathcal{B}^X \backslash \{\varnothing\}$ the equality $\tau(B) = \cup\{\tau(\{x\}) : x \in B\}$. Analogously, for repointed b-convergence spaces $(X, \mathcal{B}^X, \tau)$ we get for each $B \in \mathcal{B}^X \backslash \{\varnothing\}$ the equality $\tau(B) = \cap\{\tau(\{x\}) : x \in B\}$. The operators τ_3, τ_4, τ_5 in 2.2(i) and τ_q in 2.6(ii), respectively are set-pointed. The operator τ_1 in 2.2(i) is pointed, whereas the operators τ_2 in 2.2(ii) and τ_p in 2.6(iv), respectively are repointed.

3 Some important isomorphisms

Equiform b-convergence is playing an *important* role by considering problems of a *uniform* nature. So it is possible to describe uniform convergence or more *generally* preuniform convergence by its corresponding b-convergence. In fact, the following Theorem gives an answer in this direction. Let us denote by ***SATeb-CONV*** the full subcategory of ***eb-CONV*** consisting of the related saturated objects.

Theorem 3.1. The category ***PUCONV*** of preuniform convergence spaces and uniformly continuous maps is isomorphic to the category ***SATeb-CONV***.

Proof. We refer to 2.6(i) and *conversely* consider for a saturated equiform b-convergence space $(X, \mathcal{B}^X, \mu)$ the preuniform convergence space $(X, L\mu)$ by setting $L_\mu := \mu(X)$. □

Remark 3.2. On the *other* hand we also note that ***PUCONV*** is isomorphic to the full subcategory ***DISeb-CONV*** of ***eb-CONV***, whose objects are the discrete equiform b-convergence spaces. This can be seen as follows: For a preuniform convergence space (X, J) we consider the discrete equiform b-convergence space $(X, \mathcal{D}^X, \tau_J)$ by setting: $\tau_J(\emptyset) := \{\underline{P}(X \times X)\}$ and $\tau_J(\{x\}) := J \forall x \in X$. Conversely, we put for $(X, \mathcal{D}^X, \mu)$: $L_\mu := \{\mathcal{U} \in$

$\boldsymbol{FIL}(X \times X) : \exists x \in X \mathcal{U} \in \mu(\{x\})\}$. Then (X, L_μ) defines a preuniform convergence space. The above assignments determine the asserted isomorphism. As a corollary then we can resume that the categories ***SATeb-CONV*** and ***DISeb-CONV*** are isomorphic, too.

Definitions 3.3. For a b-convergence space $(X, \mathcal{B}^X, \tau)$ let $B \in \mathcal{B}^X \setminus \{\varnothing\}$. $\mathcal{C} \in \boldsymbol{FIL}(X)$ is called *B-Cauchyfilter* (in τ) iff it satisfies

(cf) $\mathcal{C} \times \mathcal{C} \in \tau(B)$. Then, a b-convergence space $(X, \mathcal{B}^X, \tau)$ is called *b-Cauchy space* iff it satisfies

(cau) $B \in \mathcal{B}^X \setminus \{\varnothing\}$ and $\mathcal{U} \in \tau(B)$ imply the existence of a B-Cauchyfilter $\mathcal{C} \in \boldsymbol{FIL}(X)$ with $\mathcal{C} \times \mathcal{C} \subset \mathcal{U}$.

By ***b-CAU*** we denote the full subcategory of ***b-CONV***, whose objects are the b-Cauchy spaces and additionally by ***SETb-CAU*** its full subcategory of set-pointed b-Cauchy spaces.

Theorem 3.4. The category ***SETCONV*** of set-convergence spaces and continuous maps is isomorphic to the category ***SETb-CAU***.

Proof. Taking into account 2.6(ii) we conversely put for a set-pointed b-Cauchy space $(X, \mathcal{B}^X, \mu) : \mathcal{F} p_\mu \emptyset$ iff $\mathcal{F} = \underline{P}X$ and $\mathcal{F} p_\mu B$ iff $\mathcal{F} \times \mathcal{F} \in \mu(B)$, otherwise. Then the above assignment define us the proposed isomorphism. Also note, that for set-convergence spaces $(X, \mathcal{B}^X, q)$, $(Y, \mathcal{B}^Y, p)$ a map f from $X \longrightarrow Y$ is continuous iff f is b-uniformly continuous between the related b-Cauchy spaces. □

Theorem 3.5. The category ***FIL*** of filter spaces and related maps is isomorphic to the full subcategory ***SATeb-Cau*** of ***b-CAU***, whose objects are the saturated equiform b-Cauchy spaces.

Proof. Taking into account 2.6(iii) we conversely set for a saturated equiform b-Cauchy space $(X, \mathcal{B}^X, \mu)$

$$\phi_\mu := \{\mathcal{F} \in \boldsymbol{FIL}(X) : \mathcal{F} \times \mathcal{F} \in \mu(X)\}.$$

Then the remaining statements are easily to verify. □

Remark 3.6. Analogously to remark 2.3 we also keep hold that ***FIL*** is isomorphic to the full subcategory ***DISeb-CAU*** of ***eb-CAU***, whose objects are the discrete equiform b-Cauchy spaces. (Hereby ***eb-CAU*** denotes the full subcategory of ***b-CAU***, whose objects are the equiform b-Cauchy spaces).
Thus, in *summary* we get the statement that ***SATeb-CAU*** and ***DISeb-CAU*** are isomorphic, too.

Definition 3.7. A repointed b-convergence space $(X, \mathcal{B}^X, \tau)$ is called *b-filter space* iff τ satisfies

(fil) $B \in \mathcal{B}^X \backslash \{\varnothing\}$ and $\mathcal{U} \in \tau(B)$ imply $\forall x \in B \exists \mathcal{F} \in \boldsymbol{FIL}(X)(\mathcal{F} \times \mathcal{F} \in \tau(\{x\})$ and $\mathcal{F} \times \mathcal{F} \subset \mathcal{U})$.

By ***b-FIL*** we denote the corresponding full subcategory of ***rpb-CONV***.

Theorem 3.8. The category ***GCONV*** is isomorphic to the full subcategory ***SATb-FIL*** of ***b-FIL***, whose objects are the saturated b-filter spaces.

Proof. Taking into account 2.6(iv) we conversely consider for a saturated b-filter space $(X, \mathcal{B}^X, \mu)$ the generalized convergence space (X, q_μ) by setting:

$$\mathcal{F} q_\mu x \text{ iff } \mathcal{F} \times \mathcal{F} \in \mu(\{x\}).$$

Then the remaining statements are easily to verify. □

Remark 3.9. If we denote by ***DISb-FIL*** the full subcategory of ***b-FIL***, whose objects are the discrete b-filter spaces, then ***GCONV*** and ***DISb-FIL*** are isomorphic, too, hence ***SATb-FIL*** and ***DISb-FIL*** are *essentially* the *same* up to isomorphism. The above results now allow us to consider "point convergence" on *arbitrary* B̲-sets, *not* only *restricted* to the discrete B̲-set or power set, respectively.

Supplement 3.10. As seen above antiform b-convergence seems to be an *appropriate* tool for studying *well-known* convergences in a more *broader* respectively *extended* sense! A *further* framework is given by considering the following set-convergence induced by a given antiform b-convergence space. First we will give the corresponding definition:

Definition 3.11. A set-convergence space $(X, \mathcal{B}^X, q)$ is called *reordered* iff q satisfies

(ro) $\emptyset \neq B_1 \subset B_2 \in \mathcal{B}^X$ and $\mathcal{F}qB_2$ imply $\mathcal{F}qB_1$.

We denote by ***RO-SETCONV*** the corresponding full subcategory of ***SETCONV***.

Remark 3.12. First, we note that for a given generalized convergence space (X, r) and an arbitrary boundedness $\mathcal{B}^X$, the triple $(X, \mathcal{B}^X, q_r)$ defines a reordered set-convergence space by setting:

$\mathcal{F}q_r\emptyset$ iff $\mathcal{F} = \underline{P}X$ and

$\mathcal{F}q_rB$ iff $\forall x \in B\mathcal{F}rx$, otherwise.

On the other hand let $(X, \mathcal{B}^X, \tau)$ be a set-pointed antiform b-convergence space. Then we put:

$\mathcal{F}q_\tau\emptyset$ iff $\mathcal{F} = \underline{P}X$ and

$Fq_\tau B$ iff $\dot{B} \times \mathcal{F} \in \tau(B)$, otherwise.

Evidently, $(X, \mathcal{B}^X, q_\tau)$ is a reordered set-convergence space.

Definition 3.13. A set-pointed antiform b-convergence space $(X, \mathcal{B}^X, \tau)$ is called *fil-limited* iff τ satisfies

(fl) $B \in \mathcal{B}^X \setminus \{\varnothing\}$ and $\mathcal{U} \in \tau(B)$ imply $\exists B_0 \in \mathcal{B}^X \setminus \{\varnothing\} \exists \mathcal{F}_0 \in \boldsymbol{FIL}(X)$ $(B_0 \supset B, \dot{B}_0 \times \mathcal{F}_0 \subset \mathcal{U}$ and $\dot{B}_0 \times \mathcal{F}_0 \in \tau(B_0))$.

We denote by ***FLIMb-CONV*** the corresponding full subcategory of ***SETab-CONV***. (Hereby, ***SETab-CONV*** denotes the full subcategory of ***ab-CONV***, whose objects are set-pointed).

Theorem 3.14. The categories ***RO-SETCONV*** and ***FLIMb-CONV*** are isomorphic.

Proof. With respect to remark 3.12 we conversely consider for a given reordered set-convergence space $(X, \mathcal{B}^X, p)$ the fil-limited b-convergence space $(X, \mathcal{B}^X, \mu_p)$ by setting

$\mu_p(\emptyset) := \{\underline{P}(X \times X)\}$ and

$\mu_p(B) := \{\mathcal{U} : \exists \mathcal{F}_0 \in \boldsymbol{FIL}(X) \exists B_0 \in \mathcal{B}^X \setminus \{\emptyset\} \; (\overset{\bullet}{B} \times \mathcal{F}_0 \subset \mathcal{U}, B_0 \supset B$ and $\mathcal{F}_0 p \mathcal{B}_0)\}$, otherwise.

Then the above assignments define us the asserted isomorphism. Also note, that for reordered set-convergence spaces $(X, \mathcal{B}^X, q)$, $(Y, \mathcal{B}^Y, p)$ a map $f : X \longrightarrow Y$ is continuous iff f is b-uniformly continuous between the related fil-limited b-convergence spaces. □

Theorem 3.15.
FLIMb-CONV is bicoreflective in ***SETab-CONV***.

Proof. For a set-pointed antiform b-convergence space $(X, \mathcal{B}^X, \tau)$ we set:

$\tau_{fl}(\emptyset) := \{\underline{P}(X \times X)\}$ and

$\tau_{fl}(B) := \{\mathcal{U} : \exists \mathcal{F}_0 \in \boldsymbol{FIL}(X) \exists B_0 \in \mathcal{B}^X \setminus \{\emptyset\} \; \overset{\bullet}{B}_0 \times \mathcal{F}_0 \subset \mathcal{U}, B_0 \supset B$ and $\overset{\bullet}{B}_0 \times \mathcal{F}_0 \in \tau(B_0))\}$, otherwise.

Then $(X, \mathcal{B}^X, \tau_{fl})$ is a fil-limited b-convergence space and $1_X : (X, \mathcal{B}^X, \tau_{fl}) \longrightarrow (X, \mathcal{B}^X, \tau)$ the corresponding bicoreflection. It remains to prove the following one. For a given fil-limited b-convergence space $(Y, \mathcal{B}^Y, \mu)$ let $f : (Y, \mathcal{B}^Y, \mu) \longrightarrow (X, \mathcal{B}^X, \tau)$ be b-uniformly continuous then $f : (Y, \mathcal{B}^Y, \mu) \longrightarrow (X, \mathcal{B}^X, \tau_{fl})$ is b-uniformly continuous. Without restriction let $\mathcal{U} \in \mu(B)$ for $B \in \mathcal{B}^X \backslash \{\varnothing\}$, hence by hypothesis there exist $B_0 \in \mathcal{B}^X \backslash \{\varnothing\}$ and $\mathcal{F}_0 \in \boldsymbol{FIL}(Y)$ such that $B_0 \supset B$, $\dot{B}_0 \times \mathcal{F} \in \mu(B_0)$ and $\dot{B}_0 \times \mathcal{F} \subset \mathcal{U}$ are valid. Since f is b-uniformly continuous we get $(f \times f)(\dot{B}_0 \times \mathcal{F}) \in \tau(f[B_0])$. Consequently, $f(\dot{B}_0) \times f(\mathcal{F}) = f[\dot{B}] \times f(\mathcal{F}_0) \in \tau(f[B_0])$ follows. But $f[B_0] \supset f[B]$ with $f[B_0] \in \mathcal{B}^X \backslash \{\emptyset\}$, $f(\mathcal{F}_0) \in \boldsymbol{FIL}(X)$ and $(f \times f)(\dot{B}_0 \times \mathcal{F}) \subset (f \times f)(\mathcal{U})$ are true, showing that $(f \times f)(\mathcal{U}) \in \tau_{fl}(f[B])$ is valid. □

Remark 3.16. Especially, we should *here* mention the fact that ***GEN-CONV*** can be now considered as bicoreflective subcategory of ***PUCONV*** up to isomorphism.

Now, in the next section we will examine some facts concerning the relationship between basic properties of b-convergence spaces.

4 Relations between some basic b-convergence

First we note that ***DISb-CONV*** is bicoreflective in ***b-CONV***.

Proposition 4.1. ***SATb-CONV*** is bireflective in ***b-CONV***.

Proof. For a b-convergence space $(X, \mathcal{B}^X, \tau)$ we consider $(X, \underline{P}X, \tau_{sat})$ by setting $\tau_{sat}(\emptyset) := \{P(X \times X)\}$ and $\tau_{sat}(B) := \tau(B) \forall B \in \mathcal{B}^X \backslash \{\varnothing\}$ and $\tau_{sat}(B) := \cup\{\tau(\{x\}) : x \in B\}$ for each $B \in \underline{P}X \backslash \mathcal{B}^X$. Then $(X, \underline{P}X, \tau_{sat})$ is a saturated b-convergence space and $1_X : (X, \mathcal{B}^X, \tau) \longrightarrow (X, \underline{P}X, \tau_{sat})$ the corresponding bireflection. □

Theorem 4.2. The full subcategories ***ib-CONV***, ***ab-CONV*** of ***b-CONV*** are *both* bireflective in ***b-CONV***.

Proof. For a b-convergence space $(X, \mathcal{B}^X, \tau)$ we set:

(1) $\tau_i(\emptyset) := \{\underline{P}(X \times X)\}$ and $\tau_i(B) := \{\mathcal{U} \in \boldsymbol{FIL}(X \times X) : \exists B' \in \mathcal{B}^X \backslash \{\varnothing\}(B' \subset B \text{ and } \mathcal{U} \in \tau(B'))\}$, otherwise.

Then $(X, \mathcal{B}^X, \tau_i)$ is an isoform b-convergence space and $1_X : (X, \mathcal{B}^X, \tau) \longrightarrow (X, \mathcal{B}^X, \tau_i)$ the corresponding bireflection.

(2) For a b-convergence space $(X, \mathcal{B}^X, \tau)$ we set: $\tau_a(\emptyset) := \{\underline{P}(X \times X)\}$ and $\tau_a(B) := \{\mathcal{U} \in \boldsymbol{FIL}(X \times X) : \exists B' \in \mathcal{B}^X \backslash \{\varnothing\}(B \subset B' \text{ and } \mathcal{U} \in \tau(B'))\}$, otherwise.

Then $(X, \mathcal{B}^X, \tau_a)$ is an antiform b-convergence space and $1_X : (X, \mathcal{B}^X, \tau) \longrightarrow (X, \mathcal{B}^X, \tau_a)$ the corresponding bireflection. □

Remark 4.3. Since the categories ***ab-CONV*** and ***ib-CONV***, respectively are bireflective in ***b-CONV***, limits in ***ab-CONV*** respectively ***ib-CONV*** are formed as in ***b-CONV***. Since ***b-CONV*** is topological, ***ab-CONV*** and ***ib-CONV*** *both* are topological, too.

Theorem 4.4. The full subcategory ***eb-CONV*** of ***ab-CONV*** is bireflective in ***ab-CONV***.

Proof. For an antiform b-convergence space $(X, \mathcal{B}^X, \tau)$ we set: $\tau_e(\emptyset) := \{\underline{P}(X \times X)\}$ and $\tau_e(B) := \{\mathcal{U} \in \boldsymbol{FIL}(X \times X) : \mathcal{U} \in \cup\{\tau(\{x\}) : x \in X\}\}$, otherwise. Then $(X, \mathcal{B}^X, \tau_e)$ is an equiform b-convergence space, and $1_X : (X, \mathcal{B}^X, \tau) \longrightarrow (X, \mathcal{B}^X, \tau_e)$ the corresponding bireflection. □

Remark 4.5. Since ***ab-CONV*** is topological, it follows that ***eb-CONV*** is topological, too. As a corollary we note that ***PUCONV*** is a topological category, and limits in ***PUCONV*** are constructed in ***ab-CONV*** up to isomorphism.

Theorem 4.6. The full subcategory ***b-CAU*** of ***b-CONV*** is bicoreflective in ***b-CONV***.

Proof. For a b-convergence space $(X, \mathcal{B}^X, \tau)$ we set: $\tau_c(\emptyset) := \{\underline{P}(X \times X)\}$ and $\tau_c(B) := \{\mathcal{U} : \exists \mathcal{F} \in \boldsymbol{FIL}(X)(\mathcal{F} \times \mathcal{F} \subset \mathcal{U}$ and $\mathcal{F} \times \mathcal{F} \in \tau(B))\}$, otherwise. Then $(X, \mathcal{B}^X, \tau_c)$ is a b-Cauchy space and $1_X : (X, \mathcal{B}^X, \tau_c) \longrightarrow (X, \mathcal{B}^X, \tau)$ the corresponding bicoreflection. □

Remark 4.7. If the b-convergence space $(X, \mathcal{B}^X, \tau)$ is set-pointed, then $(X, \mathcal{B}^X, \tau_c)$ is set-pointed, too. Hence, by applying theorem 3.4 ***SET-CONV*** can be considered as bicoreflective subcategory of ***SETb-CONV***.

Theorem 4.8. The full subcategory ***rpb-CONV*** of ***ab-CONV*** is bireflective in ***ab-CONV***.

Proof. For an antiform b-convergence space $(X, \mathcal{B}^X, \tau)$ we set: $\tau_{rp}(\emptyset) := \{\underline{P}(X \times X)\}$ and $\tau_{rp}(B) := \{\mathcal{U} \in \boldsymbol{FIL}(X \times X) : \mathcal{U} \in \cap\{\tau(\{x\}) : x \in B\}\}$, otherwise. Then $(X, \mathcal{B}^X, \tau_{rp})$ is a repointed b-convergence space and $1_X : (X, \mathcal{B}^X, \tau) \longrightarrow (X, \mathcal{B}^X, \tau_{rp})$ the corresponding bireflection. □

Remark 4.9. Here, we point out that last theorem leads us to consider also *point-convergence* on an *arbitrary* boundedness. By applying 3.8 *classical* point-convergence turns out to be a *special* case of this *broader* concept. Moreover, in *general* equiform b-convergence also can be dealt with by paying attention to the fact that it is repointed and pointed, as well, and therefore it can be now considered on $\underline{\mathrm{B}}$-sets, not only restricted to the *discrete* or *saturated* case, respectively. At last, *suitable* set-convergences are integrated as well by applying 3.14. Thus, as *summary* we note that antiform b-convergence seems to be an *appropriate* tool for a *common* study of all these mentioned structures or spaces, respectively.

5 Categorical concepts

We recall the defining conditions for a *concrete* category C to be called *topological*, [6].

(CT$_1$) "Existence of initial structures": For any set X, any family $(X_i, T_i)_{i\in I}$ of C-objects indexed by a class I, and any family $(f_i : X \longrightarrow X_i)_{i\in I}$ of maps indexed by I, there exists an *unique* C-structure T on X, that is *initial* with respect to $(X, f_i, (X_i, T_i), I)$. I.e. for any C-object (Y, S) a function $g : Y \longrightarrow X$ is a C-morphism from (Y, S) to (X, T) iff for every $i \in I$ the composite map $f_i \circ g : Y \longrightarrow X_i$ is a C-morphism from (Y, S) to (X_i, T_i).

(CT$_2$) "Fibre smallness": For any set X the C-fibre, i.e.. the class of all C-structures on X is a set.

(CT$_3$) "Terminal separator property": For any set X with cardinality 1 there exists *precisely* one C-structure on X.

Moreover, a topological category (construct) is cartesian closed (i.e. has *natural* function space structures), provided that for any pair (A, B) of C-objects the set $\mathrm{Mor}(A, B)$ of all C-morphisms from A to B can be equipped with the structure of a C-object, denoted by $\mathrm{Pow}(A, B)$ and called *power-object* or *natural function space*, such that the following are satisfied:

(1) The evaluation map $e : A\times\ \mathrm{Pow}(A, B) \longrightarrow B$ defined by $e(a, f) := f(a)$ for each pair $(a, f) \subset A\times\ \mathrm{Pow}(A, B)$ is a C-morphism;

(2) for each C-object C and each C-morphism $f : A \times C \longrightarrow B$ the map $\hat{f} : C \longrightarrow \mathrm{Pow}(A, B)$ defined by $\hat{f}(a)(c) := f(a, c)$ is a C-morphism.

Next, we look at the definition of a so-called *b-topological extension*: By ***BTEXT*** we denote the category, whose objects $(e, \mathcal{B}^X, Y)$ are specified by topological spaces $X := (X, t_x)$ and $Y := (Y, t_Y)$ (given by closure operators), a boundedness $\mathcal{B}^X$ and a function $e : X \longrightarrow Y$ that satisfies the following conditions:

(txt$_1$) $A \subset X$ implies $t_X(A) = e^{-1}[t_Y(e[A])$, where e^{-1} denotes the *inverse* image under e;

(txt$_2$) $t_Y(e[X]) = Y$, which means that the image of X under e is *dense* in Y.

Morphisms in ***BTEXT*** have the form $(f, g) : (e, \mathcal{B}^X, Y) \longrightarrow (e', \mathcal{B}^{X'}, Y')$, where $f : X \longrightarrow X', g : Y \longrightarrow Y'$ are continuous maps such that f is bounded, and the following diagram commutes:

$$\begin{array}{ccc} X & \xrightarrow{e} & Y \\ {\scriptstyle f}\downarrow & & \downarrow{\scriptstyle g} \\ X' & \xrightarrow{e'} & Y' \end{array}$$

If $(f, g) : (e, \mathcal{B}^X, Y) \longrightarrow (e', \mathcal{B}^{X'}, Y')$ and $(f', g') : (e', \mathcal{B}^{X'}, Y') \longrightarrow (e'', \mathcal{B}^{X''}, Y'')$ are ***BTEXT***-morphisms, they can be *composed* component-wise, i.e. $(f', g') \circ (f, g) := (f' \circ f, g' \circ g)$, where "$\circ$" denotes the *composition* of maps.

Remark 5.1. Observe, that axiom (txt$_1$) in this definition is *automatically* satisfied if $e : X \longrightarrow Y$ is a *topological embedding*. Moreover, we admit an *ordinary* boundedness $\mathcal{B}^X$, which need *not* be necessary *coincide* with the power set $\underline{P}X$.

Definition 5.2. We call such an extension $(e, \mathcal{B}^X, Y)$

(i) *strict* iff $(e, \mathcal{B}^X, Y)$ satisfies the condition

 (st) $\{t_Y(e[A]) : A \subset X\}$ forms a *base* for the *closed* subsets of Y, [1];

(ii) *symmetric* iff $(e, \mathcal{B}^X, Y)$ satisfies the condition

 (sy) $x \in X$ and $y \in t_Y(\{e(x)\})$ imply $e(x) \in t_Y(\{y\})$, [2].

6 Power-objects for antiform b-convergence

Proposition 6.1. The construct ***ab-CONV*** is a topological category.

Proof. Since ***b-CONV*** is topological and ***ab-CONV*** bireflective in ***b-CONV***, thus ***ab-CONV*** is topological, too.
In order to show that ***ab-CONV*** is cartesian closed, we firstly prove that for given antiform b-convergence spaces $(X, \mathcal{B}^X, \tau_X)$, $(Y, \mathcal{B}^Y, \tau_Y)$ there is *always* a function space Y^X available, structured *strongly* enough to make the natural evaluation map $e : X \times Y^X \longrightarrow Y((x, f)$
$\longrightarrow f(x))$ b-uniformly continuous. □

Theorem 6.2. For any pair $((X, \mathcal{B}^X, \tau_X), (Y, \mathcal{B}^Y, \tau_Y))$ of antiform b-convergence spaces the set $Y^X := \{f : (X, \mathcal{B}^X, \tau_X) \longrightarrow (Y, \mathcal{B}^Y, \tau_Y)$ is b-uniformly continuous $\}$ can be supplied in a *natural* way with an antiform b-convergence such that the evaluation map is bounded and preserves uniform filters.

Proof. We define an antiform b-convergence on Y^X by setting:
$\mathcal{B}^{Y^X} := \{B^* \subset Y^X : \forall B \in \mathcal{B}^X B^*(B) \in \mathcal{B}^Y\}$ where $B^*(B) := \{f(x) : f \in B^*, x \in B\}$ and $\tau_{Y^X}(\emptyset) := \{\underline{P}(Y^X \times Y^X)\}$ with $\tau_{Y^X}(B^*) := \{\mathcal{U}^* \in \boldsymbol{FIL}(Y^X \times Y^X) : \forall B \in \mathcal{B}^X \backslash \{\varnothing\} \forall \mathcal{U} \in \tau_X(B) \mathcal{U}^*(\mathcal{U}) \in \tau_Y(B^*(B))\}$, otherwise.
Hereby, $\mathcal{U}^*(\mathcal{U})$ denotes the filter generated by $\{U^*(U) : U^* \in \mathcal{U}^*, U \in \mathcal{U}\}$ with $U^*(U) := \{(f(x), g(x)) : (f, g) \in U^* \text{ and } (x, z) \in U\}$. Evidently, $\mathcal{B}^{Y^X}$ is boundedness on Y^X. Moreover, τ_{Y^X} clearly satisfies (bc_1) and (bc_2), respectively.

to (bc_3): Let f be a b-uniformly continuous map. We have to show that $\dot{f} \times \dot{f} \in \tau_{Y^X}(\{f\})$. For $B \in \mathcal{B}^X \backslash \{\varnothing\}$ let $\mathcal{U} \in \tau_X(B)$, then we claim $\dot{f} \times \dot{f}(\mathcal{U}) \in \tau_Y(\{f\})(B) = \tau_Y(f[B])$, because the statement $(f \times f)(\mathcal{U}) \subset \dot{f} \times \dot{f}(\mathcal{U})$ is valid.

to (a): For $\emptyset \neq B_1^* \subset B_2^*$ and $B \in \mathcal{B}^X \backslash \{\varnothing\}$ we always have $B_1^*(B) \subset B_2^*(B)$, hence $\tau_Y(B_2^*(B)) \subset t_y\tau(B_1^*(B))$ follows, since by hypothesis t_Y is antiform. Consequently $\tau_{Y^X}(B_2^*) \subset \tau_{Y^X}(B_1^*)$ results. Next, we claim that the evaluation map is bounded. First we note that on the product $X \times Y^X$ the initial b-convergence $(B^{X\times Y^X}, \tau_{X\times Y^X})$ with respect to the data $(X \times Y^X, p_X, p_{Y^X}, (X, \mathcal{B}^X, \tau_X), (Y^X, \mathcal{B}^{Y^X}, \tau_{Y^X}))$ is established (also compare with basic notations 2.1, 4.1, 4.2 and 5.1, respectively). Here, p_X denotes the projection from $X \times Y^X$ to X, and p_{Y^X} denotes the projection from $X \times Y^X$ to Y^X, respectively. For $R \in \mathcal{B}^{X\times Y^X}$ we have to verify that $e[R] \in \mathcal{B}^Y$ is valid.

By hypothesis we have $p_X[R] \in \mathcal{B}^X$ and $p_{Y^X}[R] \in B^{Y^X}$. Consequently, by definition of $\mathcal{B}^{Y^X} p_{Y^X}[R]\ (p_X[R]) \in \mathcal{B}^Y$ can be deduced. So, it remains to verify that the inclusion $e[R] \subset p_{Y^X}[R](p_X[R])$ is valid. $y \in e[R]$ implies $y = e(x, f) = f(x)$ and $p_{Y^X}(x, f) = f$ result, and $f(x) \in p_{Y^X}[R](p_x[R])$ follows, showing that the statement $y \in p_{Y^X}[R](p_X[R])$ is true. By applying (b_1) $e[R] \in \mathcal{B}^Y$ immediately follows.
In an additional step we will show that the evaluation map preserves uniform filters. So let $R \in \mathcal{B}^{X\times Y^X}$ and $\hat{\mathcal{U}}$ be an element of $\tau_{X\times Y^X}(R)$, our goal is to verify $(e\times e)(\hat{\mathcal{U}}) \in \tau_Y(e[R])$. By definition of $\tau_{X\times Y^X}$ we have $(p_X \times p_X)(\hat{\mathcal{U}}) \in \tau_X(p_X[R])$ and $(p_{Y^X} \times p_{Y^X})(\hat{\mathcal{U}}) \in \tau_{Y^X}(p_{Y^X}[R])$. Then by definition of $\tau_{Y^X}(p_{Y^X}[R])$ we get $(p_{Y^X} \times p_{Y^X})(\hat{\mathcal{U}})((p_X \times p_X)(\hat{\mathcal{U}})) \in \tau_Y(p_{Y^X}[R](p_X[R])$. Since $(Y, \mathcal{B}^Y, \tau_Y)$ is antiform b-convergence space we obtain the statement that $(p_{Y^X} \times p_{Y^X})(\hat{\mathcal{U}})((p_X \times p_X)(\hat{\mathcal{U}})) \in \tau_Y(e[R])$ by taking the former proved fact into consideration. To put an end of this it remains to verify that the inclusion $(p_{Y^X} \times p_{Y^X})(\hat{\mathcal{U}})((p_X \times p_X)(\hat{\mathcal{U}})) \subset (e \times e)(\hat{\mathcal{U}})$ is valid.
$V \in (p_{Y^X} \times p_{Y^X})(\hat{\mathcal{U}})((p_X \times p_X)(\hat{\mathcal{U}}))$ implies $V \supset U^*(U)$ for some $U^* \in (p_{Y^X} \times p_{Y^X})(\hat{\mathcal{U}})$ and some $U \in (p_X \times p_X)(\hat{\mathcal{U}})$. Hence $U^* \supset (p_{Y^X} \times p_{Y^X})[\hat{U}_1]$ for some $\hat{U}_1 \in \hat{\mathcal{U}}$ and $U \supset (p_X \times p_X)[\hat{\mathcal{U}}_2]$ for some $\hat{U}_2 \in \hat{\mathcal{U}}$. Consequently $\hat{U}_1 \cap \hat{U}_2 \in \hat{\mathcal{U}}$ follows, since $\hat{\mathcal{U}}$ is filter. Now, we will show that the inclusion $(e \times e)[\hat{U}_1 \cap \hat{U}_2] \subset U^*(U)$ holds, because then $V \in (e \times e)(\hat{\mathcal{U}})$ results, concluding this proof.

Now, let $(e \times e)((x, f), (z, g))$ be given for $((x, f), (z, g)) \in \hat{U}_1 \cap \hat{U}_2$, hence $(e \times e)((x, f), (z, g)) = (e(x, f), e(z, g)) = (f(x), g(z))$. On the other hand the following statements hold, i.e. $(p_{Y^X} \times p_{Y^X})((x, f), (z, g))$ $= (p_{Y^X}(x, f), p_{Y^X}(z, g)) = (f, g)$ and $(p_X \times p_X)((x, f), (z, g)) = (p_X(x, f), p_X(z, g)) = (x, z)$. Hence both of following statements are valid, i.e. $(f, g) \in (p_{Y^X} \times p_{Y^X})[\hat{\mathcal{U}}_1]$ and $(x, z) \in (p_X \times p_X)[\hat{\mathcal{U}}_2]$, showing that $(f(x), g(z)) \in U^*(U)$ is valid, which put an end of this. □

Remark 6.3. Now, at last we will show that for given antiform b-convergence spaces $(X, \mathcal{B}^X, \tau_X)$, $(Y, \mathcal{B}^Y, \tau_Y)$ and $(Z, \mathcal{B}^Z, Y_Z)$ the antiform b-convergence on Y^X is *weak* enough to ensure that for any b-uniformly continuous map $f : (X \times Z, \mathcal{B}^{X \times Z}, \tau_{X \times Z}) \longrightarrow (Y, \mathcal{B}^Y, \tau_Y)$ the *associated* function $\hat{f} : (Z, \mathcal{B}^Z, \tau_Z) \longrightarrow (Y^X, \mathcal{B}^{Y^X}, \tau_{Y^X})$ defined by $\hat{f}(z)(x) := f(x, z)$ is also b-uniformly continuous. First, we will prove that $\hat{f}$ is bounded. So, let $B \in \mathcal{B}^Z$, we must show $\hat{f}[B] \in \mathcal{B}^{Y^X}$. In doing so, let $B_X \in \mathcal{B}^X$, then we claim $\hat{f}[B](B_X) \in \mathcal{B}^Y$. Since $p_X^{-1}[B_X] \cap p_Z^{-1}[B] \in \mathcal{B}^{X \times Z}$ we get $f[p_X^{-1}[B_X] \cap p_Z^{-1}[B]] \in \mathcal{B}^Y$ by hypothesis. Now, it remains to verify that the inclusion $\hat{f}[B](B_X) \subset f[p_X^{-1}[B_X] \cap p_Z^{-1}[B]]$ holds. $y \in \hat{f}[B](B_X)$ implies $y = \hat{f}(z)(x)$ for some $z \in B$ and for some $x \in B_X$. Hence $y = f(x, z)$ follows. But $(x, z) \subset p_X^{-1}[B_X] \cap p_Z^{-1}[B]$ is valid, and consequently $y \in f[p_X^{-1}[B_X] \cap p_Z^{-1}[B]]$ results.

Theorem 6.4. For a triple $(X, \mathcal{B}^X, \tau_X)$, $(Y, \mathcal{B}^Y, \tau_Y)$ and $(Z, \mathcal{B}^Z, \tau_Z)$ of antiform b-convergence spaces let $f : (X \times Z, \mathcal{B}^{X \times Z}, \tau_{X \times Z}) \longrightarrow (Y, \mathcal{B}^Y, \tau_Y)$ be a b-uniformly continuous map. Then the following function $\hat{f} : (Z, \mathcal{B}^Z, \tau_Z) \longrightarrow (Y^X, \mathcal{B}^{Y^X}, \tau_{Y^X})$ defined by $\hat{f}(z)(x) := f(x, z)$ for each $z \in Z$ and for each $x \in X$ is a b-uniformly continuous map.

Proof. First, let $D \in \mathcal{B}^Z \backslash \{\emptyset\}$ without restriction, we must show that $\hat{f}[D] \in \mathcal{B}^{Y^X}$, which means $\hat{f}[D](B) \in \mathcal{B}^Y$ for each $B \in \mathcal{B}^X$. Without restriction, $B \in \mathcal{B}^X \backslash \{\varnothing\}$ implies $p_X^{-1}[B] \cap p_Z^{-1} \in \mathcal{B}^{X \times Z}$, hence $f[p_X^{-1}[B] \cap$

$p_Z^{-1}[D]] \in \mathcal{B}^Y$ by hypothesis. It remains to verify that the statement $\hat{f}[D](B) \subset f[p_X^{-1}[B] \cap p_Z^{-1}[D]]$ is valid. $y \in \hat{f}[D](B)$ implies $y = \hat{f}(z)(x)$for some $z \in D$ and for some $x \in B$, respectively. Consequently, $y = f(x,z)$ by definition of $\hat{f}$. But $(x,z) \in p_X^{-1}[B] \cap p_z^{-1}[D]$ follows, which conclude this part of proof. Secondly, let $D \in \mathcal{B}^Z \backslash \{\emptyset\}$ and $\mathcal{V} \in \tau_Z(D)$, our goal is to verify that $(\hat{f} \times \hat{f})(\mathcal{V}) \in \tau_{Y^X}(\hat{f}[D])$. To this end let $B \in \mathcal{B}^X \backslash \{\varnothing\}$ and $\mathcal{U} \in \tau_X(B)$, we must show $(\hat{f} \times \hat{f})[\mathcal{V}](\mathcal{U}) \in \tau_Y(\hat{f}[D](B))$ is valid.

We put: $\mathcal{W} := \{R \subset (X \times Z) \times (X \times Z) : \exists U \in \mathcal{U} \exists V \in \mathcal{V} R \supset (p_X \times p_X)^{-1}[V]\}$. Then $\mathcal{W} \in \boldsymbol{FIL}((X \times Z) \times (X \times Z))$ holds such that $\mathcal{W} \in \tau_{X \times Z}(p_X^{-1}[B] \cap p_Z^{-1}[D])$ is valid. Now, we claim that following statements are true, i.e.

(i) $(p_X \times p_X)(\mathcal{W}) \supset \mathcal{U} \in \tau_X(B)$;

(ii) $(p_Z \times p_Z)(\mathcal{W}) \supset \mathcal{V} \in \tau_Z(B)$.

Since the inclusion $p_X[p_X^{-1}[B] \cap p_Z^{-1}[D]] \subset p_X[p_X^{-1}[B]] = B$ is valid we get $\tau_X(B) \subset \tau_X(p_X[p_X^{-1}[B] \cap p_Z^{-1}[D]])$, and analogously we have $p_Z[p_X^{-1}[B] \cap p_Z^{-1}[D]] \subset D$.
Consequently, $\mathcal{U} \in \tau_X([p_X[p_X^{-1}[B] \cap p_Z^{-1}[D]])$ and $\mathcal{V} \in \tau_Z(p_Z[p_X^{-1}[B] \cap p_Z^{-1}[D]])$ follow.

to (i): For $U \in \mathcal{U}$ let $V \in \mathcal{V}$, hence $(p_X \times p_X)^{-1}[U] \cap (p_Z \times p_Z)^{-1}[V] \subset (p_X \times p_X)[(p_X \times p_X)^{-1}[U]] = U$ is valid, and $U \in (p_X \times p_X)(\mathcal{W})$ follows which shows $\mathcal{U} \subset (p_X \times p_X)(\mathcal{W})$. Analogously, we get $\mathcal{V} \subset (p_Z \times p_Z)(\mathcal{W})$. Consequently, $\mathcal{W} \in \tau_{X \times Z}(p_X^{-1}[B] \cap p_Z^{-1}[D])$ can be deduced. By hypothesis $(f \times f)(\mathcal{W}) \in \tau_Y(f[p_X^{-1}[B] \cap p_Z^{-1}[D])$ follows. This it remains to verify that following statements are valid, i.e.

(i) $\hat{f}[D](B) \subset f[p_X^{-1}[B] \cap p_Z^{-1}[D]]$;

(ii) $(f \times f)(\mathcal{W}) \subset (\hat{f} \times \hat{f})[\mathcal{V}](\mathcal{U})$.

to (i): But this assertion holds by proving at first.

to (ii): $S \in (f \times f)(\mathcal{W})$ implies $S \supset (f \times f)[R]$ for some $R \in \mathcal{W}$. Hence, $R \supset (p_X \times p_X)^{-1}[U] \cap (p_Z \times p_Z)^{-1}[V]$ for some $U \in \mathcal{U}$ and $V \in \mathcal{V}$, respectively. But then $(\hat{f} \times \hat{f})[V] \in (\hat{f} \times \hat{f})\mathcal{V}$ is valid, resulting into $\hat{f} \times \hat{f}[V](U) \in (\hat{f} \times \hat{f})[\mathcal{V}][\mathcal{U}]$. Now, at last we have to to show that the inclusion $(\hat{f} \times \hat{f})[V](U) \subset (f \times f)[(p_X \times p_X)^{-1}[U] \cap (p_Z \times p_Z)^{-1}[V]]$ holds. Let $((\hat{f} \times \hat{f})(z_1, z_2))(x_1, x_2)$ be given for $(z_1, z_2) \in V$ and $(x_1, x_2) \in U$, hence $((\hat{f} \times \hat{f})(z_1, z_2))(x_1, x_2) = (\hat{f}(z_1), \hat{f}(z_2))(x_1, x_2) = (\hat{f}(z_1)(x_1), \hat{f}(z_2)(x_2)) = (f(x_1, z_1), f(x_2, z_2)) = (f \times f)((x_1, z_1), (x_2, z_2))$ is valid with $p_x(x_1, z_1) = x_1$ and $p_z(x_1, z_1) = z_1$ and $p_x(x_2, z_2) = x_2$ and $p_z(x_2, z_2) = z_2$. But $((x_1, z_1), (x_2, z_2)) \in (p_X \times p_X)^{-1}[U] \cap (p_Z \times p_Z)^{-1}[V]$ deliver us the proposed result.

□

Remark 6.5. To give a *summary* we note that ***ab-CONV*** is a cartesian closed topological category in which the *important* constructs ***FMER***, ***GCONV*** and ***PUCONV*** can be *uniquely* embedded up to isomorphism. Moreover, we *expressly* keep hold that these embeddings can be even *extended* to uniform convergence or point-convergence on an *arbitrary* boundedness, not *only* restricted to a power set or discrete boundedness, respectively. At last we remember that reordered set-convergence also can be dealt with. Thus, a more general *convenient* concept is created by considering antiform b-convergence.

Remark 6.6. By purely categorical arguments the following three *exponential* laws hold in ***ab-CONV***, [6].

(1) First exponential law: $X^{Y \times Z}$ is isomorphic to $(X^Y)^Z$;

(2) Second exponential law: $(\Pi_{i \in I} X_i)^Y$ is isomorphic to $\Pi_{i \in I}(X_i^Y)$;

(3) Third exponential law: $X^{\coprod_{i \in I} Y_i}$ is isomorphic to $\prod_{i \in I} X^{Y_i}$.

At last we mention that the following interesting ***distributive*** law also holds in ***ab-CONV***, i.e. $X \times \coprod_{i \in I} Y_i$ is isomorphic to $\coprod_{i \in I}(X \times Y_i)$.

Corollary 6.7. The topological construct ***eb-CONV*** is cartesian closed.

Proof. Since ***eb-CONV*** is an isomorphism closed full subcategory which is bireflective in ***ab-CONV*** and additionally *closed* under the formation of power-objects in ***ab-CONV*** we deduce that our assertion holds. By applying theorem 2.1 it follows that especially ***PUCONV*** also is cartesian closed. □

Corollary 6.8. The topological construct ***rpb-CONV*** is cartesian closed.

Proof. First let us note that ***rpb-CONV*** is an isomorphism-closed full subcategory of ***ab-CONV***. Now, we will show that it is additionally closed under the formation of power-objects in ***ab-CONV***. For repointed b-convergence spaces $(X, \mathcal{B}^X, \tau_X)$, $(Y, \mathcal{B}^Y, \tau_Y)$ we consider the power-object $(Y^X, \mathcal{B}^{Y^X}, \tau_{Y^X})$ in ***ab-CONV***. For $B^* \in \mathcal{B}^{Y^X} \backslash \{\emptyset\}$ we have to show that following inclusion holds, i.e.
$\cap\{\tau_{Y^X}(\{f\}) : f \in B^*\} \subset \tau_{Y^X}(B^*)$. Now let $\mathcal{U}^* \in \cap\{\tau_{Y^X}(\{f\}) : f \in B^*\}$, our goal is to verify $\mathcal{U}^* \in \tau_{Y^X}(B^*)$. For $B \in \mathcal{B}^X \backslash \{\varnothing\}$ let $\mathcal{U} \in \tau_X(B)$, we must show $\mathcal{U}^*(\mathcal{U}) \in \tau_Y(B^*(B))$. To this end it remains to prove that $\mathcal{U}^*(\mathcal{U}) \in \cap\{\tau_Y(\{y\}) : y \in B^*(B)\}$. $y \in B^*(B)$ implies $y = f^*(B)$ for some$f^* \in B^*$ and for some $x \in B$.
Hence, $\mathcal{U}^* \in \tau_{Y^X}(\{f^*\})$ follows showing that $\mathcal{U}^*(\mathcal{U}) \in \tau_Y(\{f^*\}(B))$ is valid. But $\tau_Y(f^*[B]) \subset \tau_Y(\{f^*(x)\}) = \tau_Y(\{y\})$ which concludes this proving. □

Conclusion 6.9. Now, coming to a certain end we offer a diagram, showing all corresponding relations of classical and new defined categories to each other.

Diagram

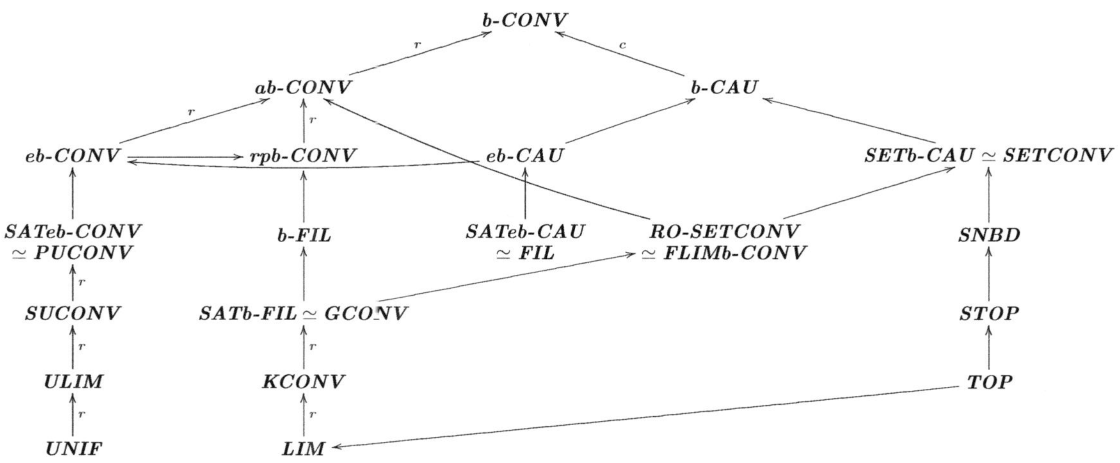

legend

SNBD	≙ Superneighborhood sp.
STOP	≙ Supertopological sp.
SUCONV	≙ Semiuniform sp.
ULIM	≙ uniform limit sp.
KCONV	≙ Kent convergence sp.
LIM	≙ limit sp.

7 Symmetric extension and related b-convergence

Now, we sill solve a *fundamental* problem in topology, roughly formulated as in the following: which b-convergence is enough *convenient* for being a candidate possessing the following properties: there exists a *dense* embedding of the space $(X, \mathcal{B}, \tau)$ into a topological one (Y, t) such that an uniform filter $\mathcal{U}$ converges to a bounded subset $B \in \mathcal{B}^X \backslash \{\varnothing\}$ iff there exists a suitable B-Cauchy filter $\mathcal{C}$ with $\mathcal{C} \times \mathcal{C} \subset \mathcal{U}$ having the *additional* property that the intersection of the closures of its sected members in Y *meet* the closure of B in Y. Note, that this problem was studied by many authors in the past but by considering more *specific* structures than b-convergences in fact. First let us list some *central* definitions of this paragraph. To handle problems of a *topological* kind it seems to be a *suitable* way by making use of isoform b-convergence in *various* manner.

Definition 7.1. Let an isoform b-convergence space $(X, \mathcal{B}^X, \tau)$ be given. Then for $B \in \mathcal{B}^X \backslash \{\varnothing\}$ a B-Cauchy filter $\mathcal{C}$ is called *B-Cauchy screen* (in τ) iff $\mathcal{C}$ satisfies the following two conditions, i.e.

(csc_1) $B \in \sec \mathcal{C}$;

(csc_2) $cl_\tau(A) \in \sec \mathcal{C}, A \subset X$ imply $A \in \sec \mathcal{C}$.

Then $(X, \mathcal{B}^X, \tau)$ is called *graded* iff τ satisfies the condition

(grd) $B \in \mathcal{B}^X \backslash \{\varnothing\}$ and $\mathcal{U} \in \tau(B)$ imply the existence of a B-Cauchy screen $\mathcal{C}$ in τ with $\mathcal{C} \times \mathcal{C} \subset \mathcal{U}$.

Now, let us denote by ***GRADb-CONV*** the full subcategory of ***ib-CONV***, whose objects are the graded b-convergence spaces. Hereby, in *general* we note that for an arbitrariy b-convergence space $(X, \mathcal{B}^X, \tau)$ $cl_\tau : \underline{P}X \longrightarrow$

$\underline{P}X$ defined by $cl_\tau(A) := \{x \in X : \{x\} \times A \in \sec \mathcal{U} \text{ for some } \mathcal{U} \in \tau(\{x\})\}$ is *closure* operator on X.

Additionally we point out, that for arbitrary b-convergence spaces $(X, \mathcal{B}^X, \tau)$, $(Y, \mathcal{B}^Y, \Gamma)$ and each b-uniformly continuous map $f : X \longrightarrow Y$ f is a continuous function from (X, cl_τ) to (Y, cl_Γ).

Example 7.2. For a surrounding space $(X, \mathcal{B}^X, \Theta)$ (see appendix 9.) or more *restricted* for a supertopological space $(X, \mathcal{B}^X, \Theta)$ we consider the triple $(X, \mathcal{B}^X, \tau_\Theta)$, where τ_Θ is defined by setting:

$\tau_\Theta(\emptyset) := \{\underline{P}(X \times X)\}$ and

$\tau_\Theta(B) := \{\mathcal{U} \in FIL(X \times X) : \Theta(B) \times \Theta(B) \subset \mathcal{U}\}$ if $B \in \mathcal{B}^X \backslash \{\varnothing\}$.
Then $(X, \mathcal{B}^X, \tau_\Theta)$ is a graded b-convergence space.

Example 7.3. For a symmetric ***BTEXT***-object $(e, \mathcal{B}^X, \tau_e)$ we obtain a graded b-convergence space $(X, \mathcal{B}^X, \tau_e)$ by setting:

$\tau_e(\emptyset) := \{\underline{P}(X \times X)\}$ and

$\tau_e(B) := \{\mathcal{U} \in FIL(X \times X) : \exists \mathcal{F} \in FIL(X)(\mathcal{F} \times \mathcal{F} \subset \mathcal{U}$ and $\cap\{t_{Y'}(e[F]) : F \in \sec \mathcal{F} \cup \{B\}\} \neq \emptyset)\}$, if $B \in \mathcal{B}^X \backslash \{\varnothing\}$.

Definitions 7.4. For a b-convergence space $(X, \mathcal{B}^X, \tau)$ τ is called

(i) *sected* iff τ has the following property

(sc) $B \in \mathcal{B}^X$ implies $\cap\{\mathcal{U} : \mathcal{U} \in \tau(B)\} \in \tau(B)$;

(ii) *topnear* iff τ satisfies the condition

(tn) $B \in \mathcal{B}^X \backslash \{\varnothing\}$ implies $\Theta_\tau(B) \times \Theta_\tau(B) \in \tau(B)$ where
$\Theta_\tau(B) := \{V \subset X : V \times V \in \cap\{\mathcal{V} : \mathcal{V} \in \tau(B)\}\}$;

(iii) *cross-near* iff τ fulfills the property

(cs) $B \in \mathcal{B}^X \backslash \{\varnothing\}, \mathcal{U} \in \tau(B)$ with $B_1 \times B_1 \in \sec \mathcal{U}$ imply $\mathcal{U} \in \tau(B_1)$

(iv) *additive* iff τ satisfies the condition

(ad) $B_1, B_2 \in \mathcal{B}^X \backslash \{\emptyset\}$ and $B_1 \cup B_2 \in \mathcal{B}^X$ imply $\tau(B_1 \cup B_2) = \tau(B_1) \cup \tau(B_2)$.

Remark 7.5. Here, we point out that for a *symmetric* set-convergence space $(X, \mathcal{B}^X, q)$ (see 2.6(ii)), where q in addition satisfies (s) $B \in \mathcal{B}^X \backslash \{\varnothing\}, \mathcal{F} q B$ with $B_1 \in \sec \mathcal{F}$ imply $\mathcal{F} q B_1$, τq is cross-near. (Note, that in the "discrete" case each symmetric set-convergence space can be ***essentially*** regarded as a symmetric generalized convergence space by satisfying $\mathcal{F} q x$ and $z \in \cap \mathcal{F}$ imply $\mathcal{F} q z$). Moreover we state, that τ_e is cross-near too, and in addition it is additive. At last let us mention that for a given surrounding space $(X, \mathcal{B}^X, \Theta)$, τ_Θ is set-pointed, sected and topnear.

Definitions 7.6. A graded b-convergence space $(X, \mathcal{B}^X, \tau)$ is called *surrounded* iff τ is set-pointed, sected and topnear as well. We denote by ***SRb-CONV*** the full subcategory of ***GRADb-CONV***, whose objects are the surrounded b-convergence spaces.

Theorem 7.7. The category ***SR*** of surrounding spaces and b-continuous maps is ismorphic to ***SRb-CONV*** (see sec. 9).

Proof. For a surrounding space $(X, \mathcal{B}^X, \Theta)$ we consider the triple $(X, \mathcal{B}^X, \tau_\Theta)$ (see example 6.2). Conversely, for a so given b-convergence space $(Y, \mathcal{B}^Y, \Gamma)$ we consider the triple $(Y, \mathcal{B}^Y, \Phi_\Gamma)$, where Φ_Γ is defined by setting:

$$\Phi_\Gamma(\emptyset) := \underline{P}Y \text{ and}$$

$$\Phi_\Gamma(B) := \{V \subset Y : V \times V \in \cap\{\mathcal{U} : \mathcal{U} \in \Gamma(B)\}, \text{ otherwise.}$$

Then $(Y, \mathcal{B}^Y, \Phi_\Gamma)$ is a surrounding space, and the corresponding *assignment* is bijective. At last, we claim that for two given surrounding spaces

$(X, \mathcal{B}^X, \Theta)$, $(Y, \mathcal{B}^Y, \Phi)$ and a bounded map $f : X \longrightarrow Y$ the following statements are equivalent, i.e.

(i) $f : (X, \mathcal{B}^X, \Theta) \longrightarrow (Y, \mathcal{B}^Y, \Phi)$ is b-continuous;

(ii) $f : (X, \mathcal{B}^X, \tau_\Theta) \longrightarrow (Y, \mathcal{B}^Y, \tau_\Phi)$ is b-uniformly continuous.

$\square$

Remark 7.8. With respect to example 7.2 we point out that now supertological spaces can be considered as *special* cases of some graded b-convergence up to isomorphism.

Definitions 7.9. A graded b-convergence space $(X, \mathcal{B}^X, \tau)$ is called *extensible* iff τ is *both* additive and cross-near. By ***EXTb-CONV*** we denote the full subcategory of ***GRADb-CONV***, whose objects are the extensible b-convergence spaces.

Lemma 7.10. For a symmetric ***BTEXT***-object $(e, \mathcal{B}^X, Y)$, $(X, \mathcal{B}^X, \tau_e)$ is an extensible b-convergence space such that t_X equals with cl_{τ_e}.

Proof. With respect to example 7.3 and remark 7.5 $(X, \mathcal{B}^X, \tau_e)$ is extensible. Now, we will verify the equality $t_X = cl_{\tau_e}$.

to "$\geq$": $x \in cl_{\tau_e}(A)$ implies the existence of a uniform filter $\mathcal{U} \in \tau_e(\{x\})$ with $\{x\} \times A \in \sec\mathcal{U}$. Hence we can find a filter $\mathcal{F} \in FIL(X)$ with $\mathcal{F} \times \mathcal{F} \subset \mathcal{U}$ and some element $y \in t_Y(\{e(x)\})$ such that $y \in \bigcap\{t_y(e[F]) : F \in \sec\mathcal{F}\}$.

By (sy) we obtain $e(x) \in t_Y(\{y\})$, and $\sec\mathcal{U} \subset \sec(\mathcal{F} \times \mathcal{F})$ deliver us that $\{x\} \times A \in \sec(\mathcal{F} \times \mathcal{F})$ holds. Consequently, $A \in \sec\mathcal{F}$ follows which implies $y \in t_Y(e[A])$. But then $e(x) \in t_Y(e[A])$ is valid, resulting into $x \in e^{-1}[t_Y(e[A])] = t_X(A)$, according to ($\text{txt}_1$).

to "$\leq$": For $x \in t_X(A)$ we set $\mathcal{F}^X := \sec\{T \subset X : x \in t_X(T)\}$, hence $\mathcal{F}^X \in \boldsymbol{FIL}(X)$ is valid. First, we will show $\mathcal{F}^X \times \mathcal{F}^X \in \tau_e(\{x\})$ and secondly that $\{x\} \times A \in \sec(\mathcal{F}^X \times \mathcal{F}^X)$. We put $y := e(x)$, hence $y \in t_Y(\{e(x)\})$. For $F \in \sec \mathcal{F}^X$ we get $x \in t_X(F)$, and $y = e(x) \in t_Y(e[F])$ follows according to (txt_1). If $D \in \mathcal{F}^X \times F^X$ then $D \supset F \times F$ for some $F \in \mathcal{F}^X$. Consequently, we obtain the statements $F \cap \{x\} \neq \emptyset \neq F \cap A$. Hence $x \in F$ and $z \in F$ for some $z \in A$ follows, showing that $(x, z) \in (\{x\} \times A) \cap (\{x\} \times F) \subset F \times F \subset D$ are valid, and $\{x\} \times A \in \sec(\mathcal{F}^X \times \mathcal{F}^X)$ results, proving the made assertion.

□

Theorem 7.11.
We obtain a functor $F : \boldsymbol{SYBTEXT} \longrightarrow \boldsymbol{EXTb\text{-}CONV}$ by setting

(a) $F(e, \mathcal{B}^X, Y) := (X, \mathcal{B}^X, \tau_e)$;

(b) $F(f, g) := f$ for a $\boldsymbol{BTEXT}$-morphism $(f, g) : (e, B^X, Y) \longrightarrow (e', \mathcal{B}^{X'}, Y')$.

(Hereby, $\boldsymbol{SYBTEXT}$ denotes the full subcategory of BTEXT, whose objects are the symmetric b-topological extensions).

Proof. We already know that the image of F lies in $\boldsymbol{EXTb\text{-}CONV}$. Now, consider a $\boldsymbol{BTEXT}$-morphism $(f, g) : (e, \mathcal{B}^X, Y) \longrightarrow$
$(e', \mathcal{B}^{X'}, Y')$. We must establish that $f : (X, \mathcal{B}^X, \tau_e) \longrightarrow (X', \mathcal{B}^{X'}, \tau_{e'})$ is a b-uniformly continuous map.
By hypothesis f is bounded. Now, let $\mathcal{U} \in \tau_e(B), B \in \mathcal{B}^X \backslash \{\varnothing\}$, our goal is to verify that $(f \times f)(\mathcal{U}) \in \tau_{e'}(f[B])$ is valid. By definition of τ_e we can choose $\mathcal{F} \in FIL(X)$ and $y \in t_Y(e[B])$ such that $y \in \cap\{t_y(e[F]) : F \in \sec \mathcal{F}\}$ and $\mathcal{F} \times \mathcal{F} \subset \mathcal{U}$ hold. Since g is continuous we get $g(y) \in t_{Y'}(g(e[B])) = t_{Y'}(e'[f[B]])$, because the corresponding diagram is commutative. We set $\mathcal{F}' := f(F)$, hence $\mathcal{F}' \in FIL(X')$ with $\mathcal{F}' \times \mathcal{F}' \subset (f \times f)(\mathcal{U})$

can be deduced. For $A \in \sec \mathcal{F}'$ we must show that $g(y) \in t_{Y'}(e'[A])$ is valid. But $A \in \sec \mathcal{F}'$ implies $f^{-1}[A] \in \sec \mathcal{F}$, hence $y \in t_Y(e[f^{-1}[A]])$ results, showing that $g(y) \in t_{Y'}(g[e[f^{-1}[A]]]) = t_{Y'}(e'[f[f^{-1}[A]]]) \subset t_{Y'}(e'[A])$ is true. Consequently, our assertion holds. □

Proposition 7.12. Let $(X, \mathcal{B}^X, \tau)$ be a graded b-convergence space, then cl_τ is a topological closure operator.

Proof. $x \in cl_\tau(cl_\tau(A))$, $A \subset X$ implies the existence of a uniform filter $\mathcal{U} \in \tau(\{x\})$ with $\{x\} \times cl_\tau(A) \in \sec \mathcal{U}$. Choose a $\{x\}$ - Cauchy screen $\mathcal{C}$ in τ with $\mathcal{C} \times \mathcal{C} \subset \mathcal{U}$ by hypothesis. Then $\mathcal{C} \times \mathcal{C} \in \tau(\{x\})$ follows, and $\sec \mathcal{U} \subset \sec(\mathcal{C} \times \mathcal{C})$ results. Consequently $\{x\} \times cl_\tau(A) \in \sec(\mathcal{C} \times \mathcal{C})$ is valid. We claim $cl_\tau(A) \in \sec \mathcal{C}$, since $F \in \mathcal{C}$ implies $F \times F \in \mathcal{C}$, and $(\{x\} \times cl_\tau(A)) \cap (F \times F) \neq \emptyset$ follows, hence $F \cap cl_\tau(A) \neq \emptyset$ is true. Since $\mathcal{C}$ is $\{x\}$-Cauchy screen in τ $A \in \sec \mathcal{C}$ results. Now, if $R \in \mathcal{C} \times \mathcal{C}$, then $R \supset F_1 \times F_1$ for some $F_1 \in \mathcal{C}$. Consequently $A \cap F_1 \neq \emptyset$, and we can choose $z \in F_1$ for some $z \in A$. But $\{x\} \times cl_\tau(A) \in \sec(\mathcal{C} \times \mathcal{C})$ implies $x \in F_1$, and $(x, z) \in (\{x\} \times A) \cap (F_1 \times F_1)$ results, showing that $(\{x\} \times A) \cap R \neq \emptyset$, and $x \in cl_\tau(A)$ is valid. □

Remark 7.13. For a b-convergence space $(X, \mathcal{B}^X, \tau)$ we can define *another* closure operator cl^τ by setting: $cl^\tau(A) := \{x \in X : \exists \mathcal{F} \in FIL(X)(\mathcal{F} \times \mathcal{F} \in \tau(\{x\})$ and $A \in \sec \mathcal{F})\}$, for each $A \subset X$. Then for a b-Cauchy space $(X, \mathcal{B}^X, \tau)$ we get for each $A \subset X$ the inclusion $cl_\tau(A) \subset cl^\tau(A)$. On the other hand we deduce the ***converse*** if considering an arbitrary symmetric b-topological extension $(e, \mathcal{B}^X, Y)$, i.e. we get for each $A \subset X$ the inclusion $cl^{\tau_e}(A) \subset cl_{\tau_e}(A)$.

Proposition 7.14. For a graded b-convergence space $(X, \mathcal{B}^X, \tau)$ and each $x \in X$ $x_\tau := \sec\{T \subset X : x \in cl_\tau(T)\}$ is $\{x\}$-Cauchy screen in τ such that $x_\tau \times x_\tau$ is *minimal* in $\tau(\{x\})$, ordered by the inclusion.

Proof. Let $x \in X$ be given. Then $x_\tau \in FIL(X)$ is valid. $x_\tau \times x_\tau \in \tau(\{x\})$, since $\dot{x} \times \dot{x} \in \tau(\{x\})$ implies the existence of a $\{x\}$-Cauchy screen $\mathcal{C}$ in τ with $\mathcal{C} \times \mathcal{C} \subset \dot{x} \times \dot{x}$. Thus $\mathcal{C} \subset \dot{x}$ follows. We will show that the inclusion $\mathcal{C} \subset x_\tau$ holds. Therefore it suffices to prove $\sec x_\tau \subset \sec \mathcal{C}$ is valid. $A \in \sec x_\tau$ and $F \in \mathcal{C}$ imply $x \in cl_\tau(A)$ and $x \in F$, showing that $cl_\tau(A) \in \sec \mathcal{C}$ is true. But by (csc_2) $A \in \sec \mathcal{C}$ follows, and $A \cap F \neq \emptyset$ results, which concludes this part of proof. Now let $\mathcal{U} \in \tau(\{x\})$ with $\mathcal{U} \subset x_\tau \times x_\tau$, hence there exists a $\{x\}$-Cauchy screen $\mathcal{C}$ in τ with $\mathcal{C} \times \mathcal{C} \subset \mathcal{U}$, hence $\mathcal{C} \subset x_\tau$ follows. On the other hand $A \in \sec \mathcal{C}$ implies $\{x\} \times A \in \sec(\mathcal{C} \times \mathcal{C})$, because $R \in \mathcal{C} \times \mathcal{C}$ implies $R \supset F \times F$ for some $F \in \mathcal{C}$.
Hence $A \cap F \neq \emptyset$ and $x \in F$ are valid. Therefore $z \in A$ for some $x \in F$ follows, showing that $(x, z) \in (\{x\} \times A) \cap (F \times F) \subset (\{x\} \times A) \cap R$. Consequently, $x \in cl_\tau(A)$ results, and $A \in \sec x_\tau$ is valid, implying $x_\tau = \mathcal{C}$. But then $\mathcal{U} = x_\tau \times x_\tau$ follows. □

8 Strict topological extension for b-convergence

In the previous section we have found a functor F : ***SYBTEXT*** $\longrightarrow$ ***EXTb-CONV***. Now we are going to introduce a related one from ***EXTb-CONV*** to ***SYBTEXT***.

Lemma 8.1. Let $(X, \mathcal{B}^X, \tau)$ be a graded b-convergence space. We put $X^* := \{\mathcal{C} \in FIL(X) : \mathcal{C}$ is B-Cauchy screen in τ for some $B \in \mathcal{B}^X \backslash \{\varnothing\}\}$, and for each $A^* \subset X^*$ we set: $t_{X^*}(A^*) := \{\mathcal{C} \in A^* : \triangle A^* \subset \sec \mathcal{C}\}$, where $\triangle A^* := \{F \subset X : \forall \mathcal{C} \in A^* \; F \in \sec \mathcal{C}\}$. Then $t_{X^*} : \underline{P}X^* \longrightarrow \underline{P}X^*$ is a topological closure operator on X^*.

Proof. Assume $t_{X^*}(\emptyset) \neq \emptyset$ and choose a B-Cauchy screen $\mathcal{C}$ in τ with $\underline{P}X = \triangle\emptyset \subset \sec \mathcal{C}$. Hence $\emptyset \in \sec \mathcal{C}$, and $\emptyset \cap F \neq \emptyset$ is valid for some $F \in \mathcal{C}$; note

that $\mathcal{C} \neq \emptyset$, and by *convention* $\underline{P}X = \Delta\emptyset$ holds. But this contradicts! Now consider $\mathcal{C} \in A^* \subset X^*$ and $F \in \Delta A^*$, hence $F \in \sec \mathcal{C}$ follows, showing that $\mathcal{C} \in t_{X^*}(A^*)$ is valid.
For $A_1^* \subset A_2^*$ and $\mathcal{C} \in t_{X^*}(A_1^*)$ we have $\Delta A_2^* \subset \Delta_1^* \subset \sec \mathcal{C}$, which shows $\mathcal{C} \in t_{X^*}(A_2^*)$.
$\mathcal{C} \in t_{X^*}(A_1^* \cup A_2^*)$ implies $\Delta(A_1^* \cup A_2^*) \subset \sec \mathcal{C}$. Assume $\mathcal{C} \notin t_{X^*}(A_1^*) \cup t_{X^*}(A_2^*)$, hence $\Delta A_1^* \not\subset \sec \mathcal{C}$ and $\Delta A_2^* \not\subset \sec \mathcal{C}$. Choose $F_1 \in \Delta A_1^* \backslash \sec \mathcal{C}$ and $F_2 \in \Delta A_2^* \backslash \sec \mathcal{C}$. We claim that both $X \backslash F_1$ and $X \backslash F_2$ belong to $\mathcal{C}$. From $X \backslash (F_1 \cup F_2) = X \backslash F_1 \cap X \backslash F_2$ we see $F_1 \cup F_2 \notin \sec \mathcal{C}$. By hypothesis choose $\mathcal{D} \in A_1^* \cup A_2^*$ with $F_1 \cup F_2 \notin \sec \mathcal{D}$. If $\mathcal{D} \in A_1^*$, then $F_1 \in \Delta A_1^*$ implies $F_1 \in \sec \mathcal{D}$, and hence $F_1 \cup F_2 \in \sec \mathcal{D}$, a contradiction. By symmetry, $\mathcal{D} \in A_2^*$ leads to a contradiction as well. Thus we have $t_{X^*}(A_1^* \cup A_2^*) = t_{X^*}(A_1^*) \cup t_{X^*}(A_2^*)$. $\mathcal{C} \in t_{X^*}(t_{X^*}(A^*))$ implies $\Delta t_{X^*}(A^*) \subset \sec \mathcal{C}$. We need to show $\Delta A^* \subset \sec \mathcal{C}$. $F \notin \sec \mathcal{C}$ implies $F \notin \sec \mathcal{D}$ for some $\mathcal{D} \in t_{X^*}(A^*)$, hence we get $\Delta A^* \subset \sec \mathcal{D}$, consequently $F \notin \Delta A^*$, which establishes the claim. □

Theorem 8.2. For graded b-convergence spaces $(X, \mathcal{B}^X, \tau)$, $(Y, \mathcal{B}^Y, \Gamma)$ let $f; X \longrightarrow Y$ be a b-uniformly continuous map. Define a function $f^* : X^* \longrightarrow Y^*$ by setting $f^*(\mathcal{C}) := \sec\{D \subset Y : f^{-1}[cl_\Gamma(D)] \in \sec \mathcal{C}\}$ for each $\mathcal{C} \in X^*$. Then the following statements are valid:

(i) f^* is a continuous map from (X^*, t_{X^*}) to (Y^*, t_{Y^*});

(ii) The composites $f^* \circ e_X$ and $e_Y \circ f$ coincide, where $e_X : X \longrightarrow X^*$ denotes the function defined by $e_X(x) := x_\tau$ for each $x \in X$.

Proof. If $\mathcal{C} \in X^*$, then $\mathcal{C} \times \mathcal{C} \in \tau(B)$ for some $B \in \mathcal{B}^X \backslash \{\varnothing\}$. We have to show that $f^*(\mathcal{C}) \times f^*(\mathcal{C}) \in \Gamma(f[B])$. By hypothesis we have $f(\mathcal{C}) \times f(\mathcal{C}) = (f \times f)(\mathcal{C} \times \mathcal{C}) \in \Gamma(f[B])$, hence there exists a $f[B]$-Cauchy screen $\mathcal{F}$ in Γ with $\mathcal{F} \times \mathcal{F} \subset f(\mathcal{C}) \times f(\mathcal{C})$. It remains to verify $\sec f^*(\mathcal{C}) \subset \sec \mathcal{F}$.

To this end, it suffices to prove $cl_\Gamma(D) \in \sec\mathcal{F}$, provided $D \in \sec f^*(\mathcal{C})$. Now in this case any $F \in \mathcal{F}$ satisfies $F \supset f[C]$ for some $C \in \mathcal{C}$, hence $f^{-1}[cl_\Gamma(D)] \in \sec\mathcal{C}$. Consequently, $f^{-1}[cl_\Gamma(D)] \cap C \neq \emptyset$. Choose $x \in C$ with $f(x) \in cl_\Gamma(D)$. Then we get $f(x) \in F$, which implies $F \cap cl_\Gamma(D) \neq \emptyset$. But now $f^*(\mathcal{C}) \times f^*(\mathcal{C})$ is a $f[B]$-Cauchy filter in Γ, and $f[B] \in f^*(\mathcal{C})$ by definition and by applying 7.1.
It remains to show that $f^*(\mathcal{C})$ satisfies (csc$_2$). But $cl_\Gamma(A) \in \sec f^*(\mathcal{C})$, $A \subset Y$ implies $f^{-1}[cl_\Gamma(cl_\Gamma(A))] \in \sec\mathcal{C}$, which shows $A \in \sec f^*(\mathcal{C})$, according to 7.12.

to (i): For $A^* \subset X^*$ we must show $f^*[t_{X^*}(A^*)] \subset t_{Y^*}(f^*[A^*])$. Choose $D \in \triangle f^*[A^*])$ with $\mathcal{D} \notin \sec f^*(\mathcal{C})$, hence $f^{-1}[cl_\Gamma(D)] \notin \sec\mathcal{C}$. By hypothesis there exists $\mathcal{A} \in A^*$ with $f^{-1}[cl_\Gamma(D)] \notin \sec\mathcal{A}$, hence $f^*(\mathcal{A}) \in f^*[A^*]$, which implies $D \in \sec f^*(\mathcal{A})$. But on the other hand, $f^{-1}[cl_\Gamma(D)] \in \sec\mathcal{A}$, which is a contradiction. Therefore $f^*(\mathcal{C}) \in cl_\Gamma(f^*[A^*])$ is valid.

to (ii): For $x \in X$ we will establish the inclusion $f^*(x_\tau) \subset f(x)_\Gamma$. To this end it satisfies to verify $\sec f(x)_\Gamma \subset \sec f^*(x_\tau)$. $T \in \sec f(x)_\Gamma$ implies $f(x) \in cl_\Gamma(T)$, hence $x \in f^{-1}[cl_\Gamma(T)] \subset cl_\tau(f^{-1}[cl_\Gamma(T)])$ follows, which means $f^{-1}[cl_\Gamma(T)] \in \sec f(x_\tau)$. But now we have $T \in \sec f^*(x_\tau)$. Since $f^*(x_\tau) \times f^*(x_\tau) \in \Gamma(\{f(x)\})$, we get that $f^*(x_\tau) \times f^*(x_\tau)$ is minimal in $\Gamma(\{f(x)\})$, ordered by the inclusion. Hence $f^*(x_\tau) \times f^*(x_\tau)$ coincides with $f(x)_\Gamma \times f(x)_\Gamma$ (see 6.14), which shows $f^*(x_\tau) = f(x)_\Gamma$, and hence $f^* \circ e_X = e_Y \circ f$, as desired.

□

Theorem 8.3.
We obtain a functor G : ***EXTb-CONV*** $\longrightarrow$ ***SYBTEXT*** by setting

(a) $G(X, \mathcal{B}^X, \tau) := (e_X, \mathcal{B}^X, X^*)$ with $X := (X, cl_\tau)$ and $X^* := (X^*, t_{X^*})$;

(b) $G(f) := (f, f^*)$ for a b-uniformly continuous map $f : (X, \mathcal{B}^X, \tau) \longrightarrow (Y, \mathcal{B}^Y, \Gamma)$.

Proof. By earlier arguments we know that cl_τ and t_{X^*} are topological closure operators on their defining sets X and X^*, respectively. Moreover, $e_X : X \longrightarrow X^*$ defined by $e_X(x) := x_\tau$ for each $x \in X$ is a function from X to X^*. Now we will establish the axioms for being a symmetric b-topological extension.

to (txt_1): For $A \subset X$ we have to show $cl_\tau(A) = e_X^{-1}[t_{X^*}(e_X[A])]$. If $x \in cl_\tau(A)$, then $\Delta e_X[A] \subset \sec x_\tau$ is valid, because $F \in \Delta e_X[A]$ implies $A \subset cl_\tau(F)$, hence $cl_\tau(A) \subset cl_\tau(F)$ results, because cl_τ is topological. Consequently, $x \in cl_\tau(F)$ is true, showing that $F \in \sec x_\tau$ is valid. In common we get $e_X(x) \in t_{X^*}(e_X[A])$, which means $x \in e_X^{-1}[t_{X^*}(e_X[A])]$ holds. Conversely, from $x \in e_X^{-1}[t_{X^*}(e_X[A])]$ we conclude $x_\tau = e_X(x) \in t_{X^*}(e_X[A])$, which implies $A \in \Delta e_X[A] \subset \sec x_\tau$, and hence $x \in cl_\tau(A)$.

to (txt_2): We must show $t_{X^*}(e_X[X]) = X^*$. For $\mathcal{C} \in X^*$ assume $\mathcal{C} \notin t_{X^*}(e_X[X])$, hence $\Delta e_X[X] \not\subset \mathcal{C}$. Choose $F \in \Delta e_X[X]$ with $F \notin \sec \mathcal{C}$. Then we have $X \subset cl_\tau(F)$. Furthermore, since $B \in \sec \mathcal{C}$ for some $B \in \mathcal{B}^X \setminus \{\varnothing\}$ $cl_\tau(F) \in \sec \mathcal{C}$ follows. But by applying (csc_2) we conclude $F \in \sec \mathcal{C}$, a contradiction.

In showing $(e_X, \mathcal{B}^X, X^*)$ is symmetric let $x \in X$ and $\mathcal{C} \in t_{X^*}\{(e_X(x)\})$, we have to verify $e_X(x) \in t_{X^*}(\{\mathcal{C}\})$.
By hypothesis we get $\Delta\{e_X(x)\} \subset \sec \mathcal{C}$, hence $x_\tau \subset \sec \mathcal{C}$, and consequently $\mathcal{C} \subset \sec x_\tau$ follows. The goal is to prove that $\Delta\{\mathcal{C}\} \subset \sec x_\tau$ is

valid. $F \in \triangle\{\mathcal{C}\}$ implies $F \in \mathcal{C}$, and $F \in \sec x_\tau$ results, showing our advisement.

At last we identify $\{t_{X^*}(e_X[A]) : A \subset X\}$ as a *base* for the *closed* subsets of X^*, which means $(e_X.\mathcal{B}^X, X^*)$ is a *strict* b-topological extension in the sense of Banaschewski [1]. If $A^* \subset X^*$ is closed in X^*, we can find some $\mathcal{C} \in X^* \backslash t_{X^*}(A^*)$, which in turn satisfies $\triangle A^* \not\subset \sec \mathcal{C}$. Hence there exists $F \in \triangle A^*$ with $F \notin \sec \mathcal{C}$. As $\mathcal{D} \in A^*$ implies $F \in \sec \mathcal{D}$ we obtain the inclusion $\triangle e_X[F] \subset \sec \mathcal{D}$ and therefore $A^* \subset t_{X^*}(e_X[F])$.

On the other hand we have $\mathcal{C} \notin t_{X^*}(e_X[F])$, since $F \notin \sec \mathcal{C}$ implies $\triangle e_X[F] \not\subset \sec \mathcal{C}$. This shows $t_{X^*}(e_X[F]) \subset A^*$ as desired. □

Theorem 8.4. Let F : ***SYBTEXT*** $\longrightarrow$ ***EXTb-CONV*** and G : ***EXTb-CONV*** $\longrightarrow$ ***SYBTEXT*** be the functors defined above. Then $F \circ G = 1_{\textbf{\textit{EXTb-CONV}}}$.

Proof. First we show that $F(G(X, \mathcal{B}^X, \tau)) = (X, \mathcal{B}^X, \tau)$ is an isomorphism for any ***EXTb-CONV***-object $(X, \mathcal{B}^X, \tau)$. Since
$F(G(X, \mathcal{B}^X, \tau)) = F(e_X, \mathcal{B}^X, X^*) = (X, \mathcal{B}^X, \tau_{e_X})$, we need to check whether $\tau_{e_X}(B) = \tau(B)$ for $B \in \mathcal{B}^X \backslash \{\varnothing\}$. Now, $\mathcal{U} \in \tau_{e_X}(B)$ implies the existence of a filter $\mathcal{F} \in FIL(X)$ with $\mathcal{F} \times \mathcal{F} \subset \mathcal{U}$ such that $\cap\{t_{Y^*}(e_X[F]) : F \in \sec \mathcal{F} \cup \{B\}\} \neq \emptyset$ holds. Choose $\mathcal{C} \in t_{X^*}(e_X[B])$ with $\mathcal{C} \in \cap\{t_{X^*}(e_X[F]) : F \in \sec \mathcal{F}\}$. Hence $\triangle e_X[B] \subset \sec \mathcal{C}$ follows, and consequently $B \in \sec \mathcal{C}$ is valid. $\mathcal{C} \times \mathcal{C} \in \tau(B_1)$ for some $B_1 \in \mathcal{B}^X \backslash \{\emptyset\}$ implies $\mathcal{C} \times \mathcal{C} \in \tau(B)$, since τ is cross-near. On the other hand we get $\sec \mathcal{F} \subset \sec \mathcal{C}$, because $F \in \sec \mathcal{F}$ implies $\mathcal{C} \in t_{X^*}(e_X[F])$. But then $\triangle e_X[F] \subset \sec \mathcal{C}$ is valid which establishes $F \in \sec \mathcal{C}$. In summary we have $\mathcal{C} \times \mathcal{C} \subset \mathcal{U}$, showing $\mathcal{U} \in \tau(B)$, as desired.
Conversely, given $\mathcal{U} \in \tau(B)$ we can find a B-Cauchy screen $\mathcal{C}$ in τ with $\mathcal{C} \times \mathcal{C} \subset \mathcal{U}$. So it remains to verify that the following two statements are valid, i.e.

(a) $\mathcal{C} \in t_{X^*}(e_X[B])$;

(b) $\mathcal{C} \in \cap\{t_{X^*}(e_X[F]) : F \in \sec \mathcal{C}\}$.

to (a): We must show $\Delta e_X[B] \subset \sec \mathcal{C}$. $F \in \Delta e_X[B]$ implies $B \subset cl_\tau(F)$, and since by hypothesis $B \in \sec \mathcal{C}$ is valid we get $F \in \sec \mathcal{C}$ according to (csc_2).

to (b): Now let $F \in \sec \mathcal{C}$, we have to verify that $\mathcal{C} \in t_{X^*}(e_X[F])$ is valid, which means $\Delta e_X[F] \subset \sec \mathcal{C}$. $A \in \Delta e_X[F]$ implies $F \subset cl_\tau(A)$, and by hypothesis as above we get $A \in \sec \mathcal{C}$, as desired.

Since $F \circ G$ maps any ***EXTb-CONV***-morphism $f : (X, \mathcal{B}^X, \tau) \longrightarrow (Y, \mathcal{B}^Y, \Gamma)$ to itself, the assertion is proved. □

Remark 8.5. The last main theorem characterize those b-convergence spaces which can be *densely* embedded into a topological one such that b-convergent uniform filters are characterized by the fact that they contain suitable bounded Cauchy filters having the property that the intersection of the closures of its sected members in this extension meets the closure of the "limit" in that construct. Note, this theorem is *not* restricted by a certain boundedness, hence it can be regarded as a *generalization* of corresponding attempts made in the past, where are studying *similar* structures on the *power set* of a given set.

Corollary 8.6. If $(X, \mathcal{B}^X, \tau)$ is *separated* that means τ satisfies

(sep) $x, z \in X$ and $x_\tau = z_\tau$ imply $x = z$, then the corresponding function $e_X : X \longrightarrow X^*$ is a topological embedding.

Corollary 8.7. Conversely, for a b-topological T_1-extension $(e, \mathcal{B}^X, Y)$, where e is a topological embedding and Y a T_1-space, then $(X, \mathcal{B}^X, \tau_e)$ is separated.

Proof. For $x, z \in X$ and $x_{\tau_e} = z_{\tau_e}$ we get $\sec x_{\tau_e} = \sec z_{\tau_e}$, hence $x \in cl_{\tau_e}(\{z\})$ follows. Consequently $\{x\} \times \{z\} \in \sec \mathcal{U}$ for some $\mathcal{U} \in \tau_e(\{x\})$. But then we can find $\mathcal{F} \in FIL(X)$ with $\mathcal{F} \times \mathcal{F} \subset \mathcal{U}$ and $\cap\{t_Y(e[F]) : F \in \sec \mathcal{F} \cup \{\{x\}\}\} \neq \emptyset$. Choose $y \in t_Y(\{e(x)\})$ with $y \in \cap\{t_Y(e[F]) : F \in \sec \mathcal{F}\}$, hence $y = e(x)$ follows, since Y is T_1-space. On the other hand $\{x\} \times \{z\} \in \sec(\mathcal{F} \times \mathcal{F})$ is valid, and $\{z\} \in \sec \mathcal{F}$ results. By hypothesis $y \in t_Y(\{e(z)\})$ can be deduced, and since Y is T_1-space we get $y = e(z)$. Consequently, $e(x) = e(z)$ follows, and $x = z$ is valid, because by hypothesis e is injective. □

9 Appendix

In this section we will now collect those topological categories which are playing a certain role in this treatise .
First, we recall the definition of a set-convergence space in the sense of Wyler [14].

Definition 9.1. Wyler defines a *set-convergence space* as a $\underline{B}$-set $\mathcal{B}^X$ equipped with a convergence relation q_X, from filters on X to bounded subsets of X, subject to the following conditions:

(sc$_1$) For $B \in \mathcal{B}^X$, the principal filter $\dot{B}$ on X converges to B, marked by $\dot{B} q_X B$;

(sc$_2$) Only the null filter on X converges to $\emptyset$, marked by:

$$\mathcal{F} q_X \emptyset \text{ imply } \mathcal{F} = \underline{P}X;$$

(sc$_3$) If $\mathcal{F} q_X B$ and $\mathcal{F} \subset \mathcal{F}_1 \in FIL(X), B \in \mathcal{B}^X$ imply $\mathcal{F}_1 q_X B$.

Then he defines a *continuous* map $f : X \longrightarrow Y$ of set-convergence spaces as a bounded map of the underlying $\underline{B}$-sets subject to the condition that

$f^{-1}(\mathcal{F}){:=}\{A \subset Y : f^{-1}[A] \in \mathcal{F}\} q_Y f[B]$ if $\mathcal{F} q_X B$, and Wyler denote by ***SETCONV*** the corresponding category of set-convergence spaces and their continuous maps, which forms a topological category.

Definition 9.2. A *supertopology* on a set X is a pair $(\mathcal{B}^X, \Theta)$, where $\mathcal{B}^X$ is a boundedness (on X) and $\Theta : \mathcal{B}^X \longrightarrow FIL(X)$ is map such that the following conditions are satisfied, i.e.

(stop$_1$) $\Theta(\emptyset) = \underline{P}X$;

(stop$_2$) $B \in \mathcal{B}^X$ and $U \in \Theta(B)$ implying $B \subset U$;

(stop$_3$) $B \in \mathcal{B}^X$ and $U \in \Theta(B)$, then there exists $V \in \Theta(B)$ such that $U \in \Theta(B_1)$ for each $B_1 \in \mathcal{B}^X$ with $B_1 \subset V$.

This definition by Tozzi + Wyler is a result by an *agreement* with Doîtchînov [13]. Then they generalize this concept to the following one, i.e. a *neighborhood structure* of a $\underline{\text{B}}$-set $\mathcal{B}^X$ is a map $\Theta : \mathcal{B}^X \longrightarrow FIL(X)$ which satisfies (stop$_1$) and (stop$_2$) and additionally the following one, i.e.

(a) $B_1 \subset B \in \mathcal{B}^X$ implies $\Theta(B) \subset \Theta(B_1)$.

Then they defined a *continuous* map $f : (X, \mathcal{B}^X, \Theta) \longrightarrow (Y, \mathcal{B}^Y, \Phi)$ of neighborhood spaces as a bounded map which satisfies:

$$B \in \mathcal{B}^X \text{ and } V \in \Phi(f[B]) \text{ imply } f^{-1}[V] \in \Theta(B).$$

By denoting ***SNBD*** the category of neighborhood spaces and their continuous maps, then ***STOP*** can be considered as full subcategory of ***SNBD*** with supertopological spaces as objects.

Definition 9.3. A *slight* specialization of a neighborhood space leads to an *intermediate* between the above and the concept of suptertopological spaces as follows: A triple $(X, \mathcal{B}^X, \Theta)$ consisting of a set X, $\mathcal{B}^X$ boundedness and a map $\Theta : \mathcal{B}^X \longrightarrow FIL(X)$ satisfying the conditions of a neighborhood structure with the additionally property of being surrounded, i.e.

(SR) $B \in \mathcal{B}^X \backslash \{\varnothing\}$ and $cl_\Theta(A) \in \sec\Theta(B)$, $A \subset X$ imply $A \in \sec\Theta(B)$, where $cl_\Theta(A) := \{x \in X : A \in \sec\Theta(\{x\})\}$ is called a *surrounding* space. Let us denote by ***SR*** the full subcategory of ***SNBD***, whose objects are the surrounding spaces.

Remark 9.4. Then obviously ***STOP*** can be nicely embedding into ***SR***. On the other hand if we consider a ***LODATO*** proximity space (X, δ) [11], and an arbitrary boundedness $\mathcal{B}^X$ on X, then the following triple $(X, \mathcal{B}^X, \Theta_\delta)$ defines a surrounding space, where Θ_δ denotes the map from $\mathcal{B}^X$ to $FIL(X)$ by setting for each $B \in \mathcal{B}^X \Theta_\delta(B) := \{A \subset X : B\bar{\delta}X\backslash A\}$. Here, $B\bar{\delta}X\backslash A$ means that B is *not* in relation to the complement of A. Next, we focus our attention to the so-called *point-convergences* or *convergence structures*, respectively, [12].

Definition 9.5. A *generalized convergence space* is a pair (X, p), where X is a set and $p \subset FIL(X) \times X$ such that the following are satisfied:

(pc_1) $(\dot{x}, x) \in p)$ for each $x \in X$;

(pc_2) $(\mathcal{F}_1, x) \in p$ whenever $(\mathcal{F}, x) \in p$ and $\mathcal{F}_1 \supset \mathcal{F}$. A map $f : (X, p_1) \longrightarrow (Y, p_2)$ between generalized convergence spaces is called *continuous* provided that $(f(\mathcal{F}), f(x)) \in p_2$ for each $(\mathcal{F}, x) \in p_1$. By ***GCONV*** we denote the correspondence category. A generalized convergence space $(\mathcal{F}, x) \in p)$ is called a ***KENT*** *convergence space* provided that the following is satisfied:

(pc_3) $(\mathcal{F} \cap \dot{x}, x) \in p$ whenever $(\mathcal{F}, x) \in p$. By ***KCONV*** we denote the full subcategory of ***GCONV***, whose objects are the KENT convergence spaces.

A generalized convergence space (X, p) is called a *limit space* provided that the following is satisfied:

(pc$_4$) $(\mathcal{F}_1 \cap \mathcal{F}_2, x) \in p$ whenever $(F_1, x) \in p$ and $(\mathcal{F}_2, x) \in p$.

We denote by $\boldsymbol{LIM}$ the full subcategory of $\boldsymbol{GCONV}$, whose objects are the limit spaces. Then *further* restrictions lead us to the category $\boldsymbol{TOP}$, but here *not* explicitely referred.

Definitions 9.6. A *filter space* is a pair (X, γ), where X is a set and γ a set of filters on X satisfying the following conditions:

(fil$_1$) $\overset{\bullet}{x} \in \gamma$ for each $x \in X$.

(fil$_2$) $\mathcal{F}_1 \in \gamma$ whenever $\mathcal{F} \in \gamma$ and $\mathcal{F} \subset \mathcal{F}_1$.

A map $f : (X, \gamma) \longrightarrow (X', \gamma')$ between filter spaces is called *Cauchy continuous* provided that $f(\mathcal{F}) \in \gamma'$ for each $\mathcal{F} \in \gamma$. We denote by *FIL* the corresponding induced category.

Definitions 9.7. A *preuniform convergence space* is a pair (X, J_X), where X is a set and J_X a set of filters on $X \times X$ such that the following are satisfied:

(uc$_1$) $\overset{\bullet}{x} \times \overset{\bullet}{x}$ belongs to J_X for each $x \in X$;

(uc$_2$) $\mathcal{U}_1 \in J_X$ whenever $\mathcal{U} \in J_X$ with $\mathcal{U} \subset \mathcal{U}_1$.

A map $f : (X, J_X) \longrightarrow (Y, J_Y)$ between preuniform convergence spaces is called *uniformly continuous* provided that $(f \times f)(\mathcal{U}) \in J_Y$ for each $\mathcal{U} \in J_X$, shortly $(f \times f)(J_X) \subset J_Y$. We denote by $\boldsymbol{PUCONV}$ the corresponding induced category. A preuniform convergence space (X, J_X) is called *semiuniform convergence space* iff J_X in addition satisfies

(uc$_3$) $\mathcal{U} \in J_X$ implies $\mathcal{U}^{-1} := \{R^{-1} : R \in \mathcal{U}\}$, where $R^{-1} := \{(z, x) : (x, z) \in R\}$.

We denote by $\boldsymbol{SUCONV}$ the full subcategory of $\boldsymbol{PUCONV}$, whose objects are the semiuniform convergence spaces.
A semiuniform convergence space (X, J_X) is called *semiuniform limit space* provided that the following is satisfied:

(uc$_4$) $\mathcal{U}_1 \in J_X$ and $\mathcal{U}_2 \in J_X$ imply $\mathcal{U}_1 \cap \mathcal{U}_2 \in J_X$.

At last, a semiuniform limit space (X, J_X) is called a uniform limit space iff J_X in addition satisfies

(uc$_5$) $\mathcal{U} \in J_X$ and $\mathcal{V} \in J_X$ imply $\mathcal{U} \circ \mathcal{V} \in J_X$ (whenever $\mathcal{U} \circ \mathcal{V}$ exists, i.e. $R \circ S := \{(x, y) : \exists z \in X$ with $(x, z) \in S$ and $(z, y) \in R\} \neq \emptyset$ for every $R \in \mathcal{U}, S \in \mathcal{V}\}$, where $\mathcal{U} \circ \mathcal{V}$ is the filter *generated* by the filter base $\{R \circ S : R \in \mathcal{U}, S \in \mathcal{V}\}$. We denote by $\boldsymbol{ULIM}$ the corresponding induced full subcategory of $\boldsymbol{SUCONV}$. To the end let us mention that a further specialization leads us to the category ***UNIF*** of uniform spaces and uniformly continuous maps, but here *not* referred.

10 References

References

[1] Banaschewski, B. Extensions of Topological Spaces. Canadian Math. Bull. 7(1964), 1–23;

[2] Bentley, H.L. Nearness spaces and extension of topological spaces. In: Studies in Topology, Academic Press, NY (1975), 47–66;

[3] Beer, G., et al. Convergence of partial maps. J. Math. Anal. Appl. 419, No 2, (2014), 1274 – 1289;

[4] Evers, K., Leseberg, D. Erratum to "A new concept of convergence space." Math. Pannonica 24 / 1 (2013), 139;

[5] Doîtchînov, D. Compactly determined extensions of topological spaces. SERDICA Bulgarice Math. Pub. 11(1985), 269–286;

[6] Herrlich, H. Cartesian closed topological categories. Math. Coll. Univ. Cape Town 9(1974), 1–16;

[7] Hogbe – Nlend, H. Bornologies and functional analysis. Amsterdam, North-Holland Pub. Co(1977);

[8] Leseberg, D. Bounded Topology : a convenient foundation for Topology. http://www.digibib.tu-bs.de/?docid=00029438, FU Berlin (2009);

[9] Leseberg, D., A new concept of convergence space. Math. Pannonica 19/2(2008)291–303, resp. http://www.digibib.tu-bs.de/?docid=00024313, TU Braunschweig (2009);

[10] Leseberg, D., On topological induced b-convergences. Top. Proc. vol. 37(2011), 293 – 313;

[11] Lodato, M.W. On topologically induced generalized proximity relations I, Proc. Amer. Math. Soc. 15. No3(1964), 417 – 422;

[12] Preuß, G., Non-symmetric Convenient Topology and its relations to Convenient Topology. Top. Proc. vol. 29(2005)595 – 611;

[13] Tozzi, A. Wyler, O., On categories of supertopological spaces. Acta Universitatis Carolinae – Mathematica et Physica 28(2), (1987), 137 – 149;

[14] Wyler O., On convergence of filters and ultrafilters to subsets. Lecture Notes in Comp. Sci. 393(1988), 340–350.

FSC
www.fsc.org
MIX
Papier aus verantwortungsvollen Quellen
Paper from responsible sources
FSC® C105338

Printed by Books on Demand GmbH, Norderstedt / Germany